CITIES AND NATURAL PROCESS

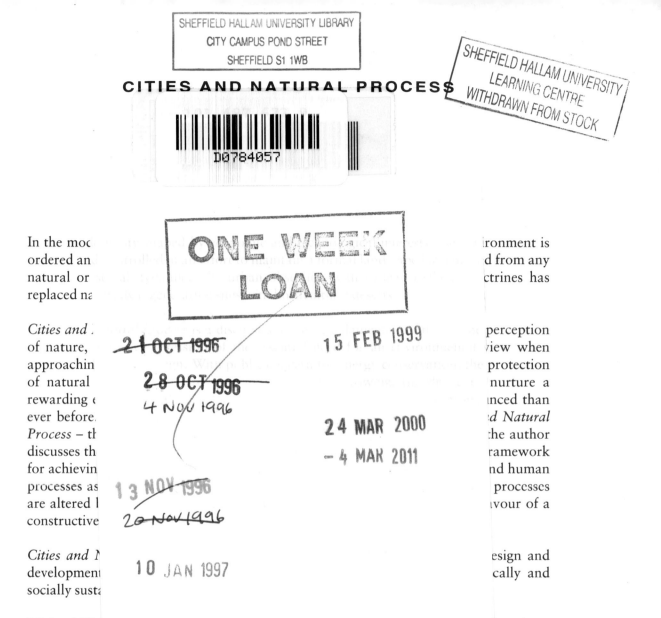

In the mo... ...ironment is
ordered an... ...d from any
natural or... ...ctrines has
replaced na...

Cities and ... ...perception
of nature, ...view when
approachin... ...protection
of natural ... ...nurture a
rewarding ... ...nced than
ever before. ...d Natural
Process – th... ...he author
discusses th... ...ramework
for achievin... ...nd human
processes as... ...processes
are altered ... ...vour of a
constructive

Cities and ... ...esign and
development... ...cally and
socially susta...

Michael Ho... ...Stansbury
Woodland N... ...Studies at
York University, Ontario, Canada. His previous books include *City Form and Natural Process* (1984) and *Out of Place* (1990).

CITIES AND NATURAL PROCESS

●

Michael Hough

London and New York

First published 1995
by Routledge
11 New Fetter Lane, London EC4P 4EE

Simultaneously published in the USA and Canada
by Routledge
29 West 35th Street, New York, NY 10001

© 1995 Michael Hough

Typeset in Sabon by
Solidus (Bristol) Limited

Printed and bound in Great Britain by
Butler and Tanner Ltd, Frome and London

British Library Cataloguing in Publication Data

A catalogue record for this book is available from the British Library

Library of Congress Cataloguing in Publication Data

Hough, Michael.
Cities and natural process/Michael Hough.
p.   cm.
Includes bibliographical references
and index.
1. City planning – Environmental aspects.
2. Urban ecology.
I. Title.
HT166.H664   1994
304.2—dc20   94–31155

ISBN 0–415–12168–X
0–415–12198–1 (pbk)

CONTENTS

•

6   Climate: Making Connections   245

PLATES

•

FIGURES

•

TABLES

•

FOREWORD

•

When the history of the decades between the 1960s and 1990s comes to be written, the attitudes expressed, and the solutions proposed and demonstrated by Michael Hough will have assumed an important place among changing world-views.

It is well to recall that in those earlier decades – often challenging the dominant received opinion – the most perceptive minds in North America were at work to uncover a new rationale for the survival of the human race. It was then – and remains among powerful groups – the dominant view that the function of humankind was and is to conquer and subdue the natural environment for current marketing goals, and exclusively to serve current occupants. To think about posterity ('What the hell has posterity done for ME?'), or to think and act constructively in behalf of generations yet unborn was then, and remains today, a minority enterprise in North American culture.

Against this tradition of grab-it-and-git, of devil-take-the-hindmost, a small but growing band of environmental designers, including Michael Hough, have been hacking away at traditional exploiters – not by personal attack, but by organizing their thoughts in professional practice, and in teaching a new form of evolutionary change.

First, some of the old superstructure had to be cleared away – the widespread grip of the 'pedigreed' tradition in landscape design; an aesthetic doctrine wrapped in the rigid forms of seventeenth- and eighteenth-century European civic designs.

Next, Hough and like-minded others found themselves demonstrating that urban life, separated from natural processes, is destructive, and that the single, exclusive use of landscape is wasteful and ill-suited to human survival. Rather, landscapes should be studied for their multiple uses, and multiple returns for least effort. As one of those constructive enterprisers, Michael Hough (expanding the footsteps of his great teacher Ian McHarg) has helped to reinvent vernacular forms and rural practices that carry a new message for city dwellers. These include water bodies which (following Chinese examples) can provide substantial food for city populations, city farms for home-town food production, natural filters for stormwaters and city forests as part of an interactive urban system. The goal is the creation of new landscapes to produce healthy life-systems, not merely to remedy urban malfunctions. On this expanded view, 'urban development becomes a participant in the workings of natural systems'. Urbanization is no longer a deadening form of desiccation.

Michael Hough expresses some of the mysterious presence of landscape energies, while peeling away some of the arcane mumbo-jumbo that has infected the dominant engineering approach to city planning and design. To this well-organized and illustrated second edition, he adds new examples from the international range of his teaching and professional practice.

Grady Clay, *Louisville, Kentucky, 28 February 1994*

PREFACE

•

Since *City Form and Natural Process* was first published in 1984, there have been some important changes in people's perceptions of nature within cities, and an increasingly sophisticated understanding of environmental issues. An awareness that natural processes and human affairs are inseparable issues is beginning to emerge. This is reflected in a commitment to environmental action that has been gathering steam over several decades. Dedicated citizens, finding new powers to effect change, are planting trees, building wetlands, bringing back native wildflowers to prairie landscapes and restoring their rivers and watersheds. A variety of agencies are restoring woodland and naturalizing roadsides and parks. If surveys are to be believed, the Times Mirror Magazines National Environment Forum, conducted in 1992, yielded the heartening conclusion that 'Americans are not willing to trade off environmental protection for economic development, despite the weak economy.' New organizations concerned with the greening of cities, ecological restoration, urban wildlife and wildflower groups have emerged, together with a host of others that are finding new sources of public support. The buzz words have also changed, reflecting the latest wisdom on the environment, from a 'Conserver' society to one concerned with 'Sustainability'. I have also learned, with one of those blinding flashes of the obvious that we like to call insight, that initiating change to the way things are done is influenced, at least initially, more by changes to deeply rooted values and traditions, than by economic imperatives. These events have made an update to this book necessary. I have, therefore, changed the title, rewritten much of it with new material and examples, while abandoning those that no longer seem pertinent to the discussion. I have also revisited the sites of the examples I have kept and provided a historical overview of how and why they have changed over a decade or more. The structure of the book has also been rearranged with Climate now appearing as the final chapter that brings the argument to a conclusion.

The task of embracing an ecological view of cities is a continually evolving process that has only begun. So, with these changes, intended to broaden and enrich the book's scope, and with some changes in focus to bring it into line with contemporary values and concerns, the fundamental issues raised in the 1984 edition still remain relevant and urgent matters that must be addressed at the end of the twentieth century, and into the next.

ACKNOWLEDGEMENTS

•

Many friends and colleagues have helped and advised me in the writing of this new edition and I am much indebted to them all. A few, however, must be acknowledged with particular gratitude. David Crombie, David Carter and Suzanne Barrett of the Waterfront Regeneration Trust in Toronto for the inspiration they have provided through their work, in which I have been privileged to participate; Emanuel Carter for his review of the first edition and valuable advice; and Tristan Palmer for his persistence in getting this book published. My thanks too to my colleagues at the Faculty of Environmental Studies at York University, for continuing to provide the stimulating academic environment that helped shape the first edition; to my colleagues at Hough Stansbury Woodland Naylor Dance with whom I have collaborated for many years in exploring and realizing creative ideas; and to my secretary Michelle Bruton for her patient help in putting the manuscript together.

CREDITS

•

INTRODUCTION

•

This book is about natural process, cities and design. Its overall thesis is that the traditional design values that have shaped the physical landscape of our cities have contributed little to their environmental health, or to their success as civilizing, enriching places to live in. My purpose is to find new and constructive ways of looking at the physical environment of cities. An alternative basis for urban landscape form is urgently needed, one that is in tune with, and will support, a growing environmental awareness of cities and nature. This book has been written with two purposes in mind. The first is to offer a conceptual, philosophical base on which urban design can rest, one that has received, until recently, scant attention in the literature. The second is to illustrate with examples drawn from real life how the practical application of theory is relevant and useful to the urban designer. The book deals with five general areas of concern.

First, there is the alienation of urban society from environmental values that embrace both the city and the land. The technology that sustains the modern city has touched every corner of human life, every landscape and wilderness, no matter how remote, and reinforces this isolation. This fact was forcefully demonstrated to me on a journey I made to the Hudson's Bay lowlands many years ago. I searched for an image of the great unspoiled Canadian wilderness, free from the sights, sounds and pressures of the urbanized south. Armed with hip waders and binoculars, I spent many days tramping through a landscape of water, muskeg and granite boulders, the dome of the sky creating a feeling of extraordinary wildness and beauty. Forget the dense clouds of hungry mosquitoes and blackfly (known as the scourge of the north), the discomforts of permanently wet boots; here was wild nature. Yet one day, a pink object, lying in the tangle of sedges at the edge of a pond, caught my eye. It was the rubber nipple from a baby's bottle, abandoned there by a passing group of Inuit. The rude shock of this relic of society, so alien to the environment around me, brought me abruptly back to reality. The incident was a powerful reminder that the products of the city are everywhere, even in the remoteness of the Hudson Bay lowlands. The nipple, my hip waders and binoculars, and the fact that I was there, transported by plane hundreds of kilometres from Toronto, verified that urbanism is a pervasive fact of life. This is so, not only for the white Canadian, but for the native individual today who uses the white Canadian's canned foods and machinery, hunts with a rifle, travels by skidoo and lives in permanent northern communities. The perceptual distinction between city and countryside has been a root cause of many social and environmental conflicts and the lack of attention to the environment of cities where most problems begin.

Second, little attention has been paid to understanding the natural processes that have contributed to the physical form of the city and which in turn have been altered by it. In the presence of plentiful energy, the urban environment has been shaped by a technology whose goals are economic rather than environmental or social. The explosive growth of urban areas since the Second World War has brought about fundamental changes, not only to the physical landscape, but to people's perceptions of land and environment. An affluent and mobile urban society now takes refuge in the countryside in search of fresh air and natural surroundings that are perceived to be denied at home. Consequently, unsustainable pressures are often placed on originally self-perpetuating ecosystems. The advancing city replaces complex communities of woods, fields and streams with biologically sterile environments that are neither socially nor visually enriching. At the same time, efforts to reclaim derelict land replaces naturally regenerated sites for new horticultural deserts, perpetuating the very conditions that they intend to cure.

Third, there are issues that concern urban processes. Apart from the vast areas of inefficiently used land in city centres and urban fringes, there are enormous water, energy and nutrient resources that are the by-products of urban drainage, sewage disposal and other functions of city processes. Having no perceived value, these contribute instead to the pollution loads of an overstressed environment.

Fourth, there are questions of aesthetic values from which the city's formal landscape has evolved. These values have little connection with the dynamics of natural process and lead to misplaced priorities. Recreation and amenity are seen as the exclusive function of urban spaces. Engineering and horticultural science, not ecological processes, determine their development, form and management. At the same time, another landscape, the fortuitous product of natural and cultural forces, flourishes without care and attention. These two landscapes symbolize a fundamental conflict of values in the perceptions of nature: the desire to nurture the one and suppress the other in a perpetual and costly struggle to maintain order and control. At a time when natural processes are being recognized as central to planning, they have, until recently been virtually ignored in the urban context. It is clear that the conventional framework for the design of cities must be re-examined. In-built assumptions about traditional priorities and standards must give way to more unconventional approaches. How can a recognition of the essential relationships between natural processes, people and economy provide clues to the shaping of cities? What are the role and function of parks and open spaces in creating healthy and dynamic places? How can the realities of multi-cultural communities in cities be recognized in design and as a relevant basis for an urban aesthetic? What are the implications of evolving citizen initiatives on urban form and action?

Fifth, there are questions of environmental values and perceptions and of how we respond to the environments around us. If it can be shown that there are cheaper, more socially valuable ways of shaping urban landscapes than has traditionally been the norm, then we have a realistic and practical basis for action. The biologist and city planner Patrick Geddes once remarked that 'civics as an art has to do not with imagining an impossible no-place where all is well, but making the most and the best of each and every place, especially in the city in which we live'.[1] So utopian ideals of the perfect city set in bucolic landscapes that were once the fashion in planning and architectural philosophy are not relevant to our concerns. This book is concerned with approaches to urban design that focus on existing cities, since it is here that the opportunities lie and where the effort must be made.

The forthcoming chapters explore these and other issues in further detail. Chapter 1 suggests the design framework within which urban natural processes should be examined. It describes the general character and evolution of the urban landscape in both the pre-industrial and modern city, and examines the constraints of energy, environment and social necessity that have helped shape their form, character and use. The attitudes and values that pervade urban life are reviewed in terms of the environmental problems they have generated and the opportunities that exist for creating a rational basis for design. Some basic principles that derive from the application of ecology to the design process are suggested. These become the frame of reference for subsequent discussions of the city's physical and social environment.

The chapters on water, plants, wildlife and city farming (Chapters 2, 3, 4 and 5) examine the various components of the natural and human environment in several ways: first, as they operate as natural systems, or in balance with nature; second, how they are affected or changed by urban processes and the attitudes and cultural values that these changes have engendered. Some alternative values based on ecological insights are suggested that would tip the balance in favour of a constructive relationship to the urban environment. Such a change also reveals opportunities instead of problems and substitutes economy for high cost. Practical examples of opportunities that are often unrecognized, but occur everywhere in the city, serve to illustrate the potential that exists for beneficial change. The implications of the ecological view on urban form are then examined through various case-studies as the foundation for a coherent philosophy of design. Chapter 6 on climate connects the various themes of the book and develops an integrated view that includes both an ecological and behavioural framework for urban design in the local scale of the city, and in its larger world context.

1

URBAN ECOLOGY:
A BASIS FOR SHAPING CITIES

•

As we approach the twenty-first century, the environmental concerns and values that began to take root in the 1960s have brought into sharp focus an acute awareness of the earth's fragility as a natural system. We have begun to understand human beings as biological creatures immersed in vital ecological relationships within the biosphere; with the need to live within its limits, sharing the planet with non-human life. These perceptions are leading us to a view that there must be a transition from a society preoccupied with consumerism and exploitation, to one that gives priority to a more sustainable future. Notwithstanding the Bruntland Commission's essentially homocentric (and perhaps necessary) interpretation of sustainability as 'meeting the needs of the present without compromising the ability of future generations to meet their own',[1] there is the ultimate need for an ethic that recognizes the interdependence of all life forms and the maintenance of biological diversity. Sustainability, therefore, becomes everyone's concern. When we consider a world population projected to be 10 billion by the year 2025, 4.5 billion people in developing countries estimated to be living in urban areas by the end of the century, and the massive impacts of human activities on world ecosystems, then it is clear that the links between nature, cities and sustainability have profound implications for survival. McHarg, Lewis and others concerned with bringing together nature and human habitat, have shown eloquently that the processes that shape the land and the limitless complexity of life forms that have evolved over evolutionary time, provide the indispensable basis for shaping human settlements. The dependence of one life process on another, the interconnected development of living and physical processes of earth, climate, water, plants and animals, the continuous transformation and recycling of living and non-living materials, these are the elements of the self-perpetuating biosphere that sustain life on earth and which give rise to the physical landscape. They become central determinants that shape all human activities on the land.

If we consider the urban landscape in this context, we find some fundamental contradictions and paradoxes in the way the city and the larger environment are perceived. In a world increasingly concerned with the problems of a deteriorating environment, be they energy, pollution, vanishing plants, animals, natural or productive landscapes, there remains a marked propensity to bypass the environment most people live in – the city itself. Conventional wisdom has traditionally regarded the modern city as the product of cheap energy, economic forces, high technology and a view of nature that is under control. The underlying disciplines that have shaped the city have little to do with the natural sciences or ecological values. If urban design can be described as that art and science dedicated to enhancing the quality of the physical environment in cities, to providing civilizing and enriching places for the people who live in them, then the current basis for urban form must be re-examined. It is necessary to rediscover, through the insights that the natural sciences provide, the nature of the familiar places we live in. Thus the basic premise on which this book rests is twofold. First, that an environmental view is an essential component of the economic, engineering, political and design processes that shape cities. The often unrecognized natural processes occurring within cities provide us with an alternative basis for their evolution and form. Second, that the problems facing the larger regional context of the countryside have their roots in cities and that solutions must, therefore, also be sought there. Thus, the task is one of linking the concept of urbanism with nature. My purpose in this chapter is to identify current problems and examine the structure and principles for design that are central to this point of view.

THE CONTRADICTION OF VALUES

Towns and cities are perceived largely through their exterior environment. The average urban dweller going about the business of daily living will experience the city through its patterns of streets and pedestrian ways, shopping places, civic squares, parks and gardens and residential areas. However, there is another generally ignored landscape lying beneath the surface of the city's public places and thoroughfares. It is the landscape of industry, railways, public utilities, vacant lands, urban expressway interchanges, abandoned mining lands and waterfronts. Thus two landscapes exist side by side in cities. The first is the nurtured 'pedigreed'[2] landscape of mown turf, flowerbeds, trees, fountains and planned places everywhere that have traditionally been the focus of civic design. Its basis for form rests in the formal design doctrine and aesthetic priorities of established convention. Its survival is dependent on high energy inputs, engineering and horticultural technology. Its image is that of the design solution independent of place: it can be found everywhere from Washington DC to Jakarta, Indonesia; from the city centre to the outlying suburbs. The second is the fortuitous landscape of naturalized urban plants and flooded places left after rain, that may be found in the forgotten places of the city. Urban 'weeds' emerge through cracks and gratings in the pavement, on rooftops, walls, poorly drained industrial sites or wherever a foothold can be gained. They provide shade and flowering groundcovers and wildlife habitat at no cost or care and against all the odds of gasoline fumes, sterile or contaminated soils, trampling and maintenance men. There

Plate 1.1 *Two urban landscapes*

A formally landscaped boulevard, and an abandoned waterfront site. Which is the derelict site?

a *Has four or five species of plants and supports no wildlife*

b *Supports over 400 species of plants and is visited by 290 species of birds*

Plate 1.2 *The built urban vernacular*

Ethnic neighbourhoods have similar underlying characteristics as the abandoned site. Both have evolved in response to minimum intervention from authority

is also a humanized landscape hidden away in the back alleys, rooftops and backyards of many an ethnic neighbourhood that can be described as the product of spontaneous cultural forces. It is here that one may find a rich variety of flourishing gardens and brightly painted houses. The turfed front yard of the well-to-do neighbourhood gives way to sunflowers, daisies, vegetable gardens, intricate fences, ornaments and religious icons of every conceivable variety, expressing rich cultural traditions, and the imperatives of necessity. The forces that shape the built vernacular, in fact, have remarkable similarities to the fortuitous landscape. Both have evolved in response to minimum interference from authority.

These two contrasting landscapes, the formalistic and the natural, the pedigreed and the vernacular, symbolize an inherent conflict of environmental values. The first has little connection with the dynamics of natural process. Yet it has traditionally been held in high public value as an expression of care, aesthetic value and civic spirit. The second represents the vitality of altered but none the less functioning natural and social processes at work in the city. Yet, it is regarded as a derelict wasteland in need of urban renewal, the disorderly shambles of the poorer parts of town. If we make the not unreasonable assumption that diversity is ecologically and socially necessary to the health and quality of urban life, then we must question the values that have determined the image of nature in cities. A comparison between the plants and animals present in a regenerating vacant lot, and those present in a landscaped residential front yard, or city park, reveals that the vacant lot generally has far greater floral and faunal diversity than the lawn or city park. Yet all efforts are directed towards nurturing the latter and suppressing the former. The reclamation of 'derelict' areas, or the creation of new development at the city's edge where the native and cultural landscape is replaced by a cultivated one, involves reducing diversity, rather than enhancing it. The question that arises, therefore, is this: which are the derelict sites in the city requiring rehabilitation? Those fortuitous and often ecologically diverse landscapes representing urban natural forces at work, or the formalized landscapes created by design?

It is my contention that the formal city landscape imposed over an original natural diversity is the one in need of rehabilitation. While it will be obvious that such a landscape has a time-honoured place in the city, its *universal* application for the making of urban places is the most persuasive argument for considering it as a derelict landscape. Other paradoxes become apparent when we apply ecological insights to our observation of the city environment.

- Attitudes and perceptions of the environment expressed in town planning since the Renaissance, have, with some exceptions, been more concerned with utopian ideals than with natural process as determinants of urban form. Examples of cities and institutions all over the world attest to the aesthetic and cultural baggage of a past era, transported to hostile climatic environments and wholly inappropriate to them. Cheap fossil fuel together with a misguided sense of civic pride, or expressions of power and wealth, has enabled the inorganic structure of planning theory to persist and maintain the illusion that the creation of benign outdoor climates has little relevance to urban development.
- Traditional storm drainage systems, the conventional method of solving the problem of keeping the city's paved surfaces free of water, have until recently been unquestioned. As the established vocabulary of engineering, water drains to the catchbasin. But the benefits of 'good design' – well-drained streets and civic spaces – is paid for by the environmental costs of eroded streams, flooding and impairment of water quality in downstream watercourses.
- Sewage disposal systems are seen as an engineering rather than a biological solution

to the ultimate larger problem of eutrophication of water bodies and wasted resources. We have the paradox of the city as the centre for enormous concentrations of nutrient energy, while urban soils remain sterile and non-productive.

• The notions of humanity and nature have long been understood to be separate issues. Such a dichotomy has had profound influences on the way people have thought about themselves: the cities where people live and the non-urban regions beyond the city where nature lives. In the unique culture from which the disciplines of intervention spring – engineering, building, planning and design – this perceived separation has also had a profound effect on the control, not only of nature, but of human behaviour. Thus the nature of pedigreed design has had little time or understanding for the innate forces that shape human environments, or the peculiar needs of multi-cultural communities that are the norm in most cities today.

The resolution of these contradictions must be found in an ecological view that encompasses the total urban landscape and the people who live there. This includes the unstructured spatial and social environments that are not currently seen to contribute to the city's civic image as well as those that do. To explore this point of view further we must examine the legacy that pre-industrial as well as modern cities have left.

VERNACULAR LANDSCAPES
AND THE INVESTMENT IN NATURE

As I have already indicated, towns and cities are perceived through their exterior environment. The way urban spaces are used is a measure of the attitudes and values of people in relation to the places they live in. While the pressures of modern times have radically altered the old pre-industrial settlements, many still maintain something of their original character. They have a quality that creates a variety of unmistakable impressions. One is that they are places for walking. There is a human scale to the streets as they wind their way informally but purposefully through the town, sometimes narrow and enclosing, sometimes opening to vistas of squares, marketplaces, common green, churchyards. Another is the countryside that suddenly opens up over rooftops, or at the end of a street; fields and woods forming a hard and defined edge between urban and non-urban places. Another is their work. There is business in the marketplace and in the street, livestock graze on the common and in the churchyard, crops ripen in the fields. Towns like this aptly but superficially fit the postcard image of picturesque quaintness that has long attracted tourists. But at heart their shape, the arrangements of their open spaces, their relationship to the countryside and harmonious siting, are the historical product of economic and social forces and the physical constraints of the land. Early

Plate 1.3 *Expressions of the vernacular urban landscape*

a *Visual connections to the countryside*

b *Places for walking*

c *Negotiations at the local sheep market*

settlement, as Mumford has pointed out, could not grow beyond the limits of its water supply and food sources until better transportation and a more sophisticated administration could evolve.[3] This early association with food production maintained a connection between the country and the city in some form until the Industrial Revolution. For instance, fruit and vegetables consumed in New York and Paris came from market gardens whose soils had been enriched with night refuse. A good part of the population had private gardens and practised rural occupations within the city, as American towns did up to 1890. Mumford also notes that the amount of usable open space within medieval cities throughout their existence was greater per head of population than any later form of city.[4]

From a design viewpoint, the most significant impression one receives of the pre-industrial town is that it made the most of what it had within the means and technology available. Since it was built and operated on solar power, it was limited by what stored energy was available from organic materials: running water and direct sunlight. Variations of climate, topography, agricultural soils and water supply shaped its form. Open spaces were functional, producing a variety of fruit and vegetables; the common and churchyard provided grass and were kept trim by livestock – a practice still kept up by many towns and cities in Europe. Groupings of houses around greens and courtyards were arranged on the basis of functional necessity to conserve heat, minimize winds and to provide sunlight and space.

Books on architecture have marvelled at the subtle sequence of spaces, the proportions of squares, the powerful architectural statement of important buildings and other aesthetic urban qualities. Rudofsky has shown how history has been preoccupied with 'pedigreed' styles of building based on formal rules of design and commemorating power and wealth.[5] It excludes 'vernacular' building which responds to the environment, to social, economic and functional necessity. We can draw relevant parallels in the landscape. The literature on historical landscapes deals almost exclusively with the development of the artistic philosophies that produced the great parks and gardens of the times, from which much of our urban park tradition can be traced. It ignores the working vernacular landscape of town and country, created out of necessity and poverty, that symbolized the investment in nature and land. But it is these that hold crucial lessons for us today in our search for a relevant basis for urban form.

ENERGY LANDSCAPES AND THE CONTEMPORARY CITY

The patterns of space in the modern city are the product of market forces, transportation systems and design ideologies that are radically different from the older city building tradition. Visually, the advent of the tower block and the freeway of post-war development has created a landscape of extraordinary scale, relating more to the automobile than to the pedestrian. The dimension of speed creates new and powerful experiences of the city. It is appreciated from the highway in broad outline – a series of changing images and fleeting impressions from a distance that are visually stimulating but sensually remote. Buildings tend to float in a sea of space rather than containing it. Tower blocks rise from open plazas which are often windswept and shadowed in cold climates, or sunbaked in hot ones. Economic forces have created a landscape of uncontained plazas, parking lots, vehicular thoroughfares, highway interchanges and vacant sites. It is a landscape sterilized by ineffective use, by a lack of co-ordination of various public and private agencies that control it and, on the city's edges, by restrictive zoning that inhibits human interaction and organic neighbourhood evolution. Aesthetic conventions and values have created a development landscape of parks, playgrounds, recreation spaces and front yards, whose character rests on a universal application of cultivated turf, asphalt and chainlink fences occasionally punctuated by an ornamental tree or exotic shrub. Of all the varied impressions that come to mind as one looks at the modern city, there are, maybe, four that reveal the most about the subject at hand: a lack of visual connections to the countryside, the use of urban parks solely for leisure, the mutually exclusive nature of the relationship between town and countryside, and the abundant use of energy.

Visual connections to the countryside

The view to the countryside from the town, the symbol of the pre-industrial dependence on the land, has disappeared. The traditional relationship between city and farmlands has been replaced by an industrialized agriculture that has no direct connections with the local place. Land that once produced crops and livestock is now more valuable as real estate. The massive exploitation of the earth that has permitted the growth of the urban region, has only recently, in the last decades of the twentieth century, been seen as a relevant problem in North America.[6] So the countryside immediately surrounding the city, known as the urban shadow, is the object of land speculation and sporadic development, denying planning solutions and perpetuating an unproductive landscape.

Parks are for recreation

Recreation has become the exclusive land use for the city's public open spaces. The migration of people from the countryside to the cities that began in the Industrial Revolution did more than create poverty and slums. The skills and knowledge of traditional patterns of rural life were replaced by the living and working patterns of the city. The psychological and physical separation between urban and rural environments widened as cities grew larger, more industrialized and more remote from the rural areas with which originally they had been connected. The urban park had an entirely different purpose from the countryside it replaced. The crops, orchards and livestock that had been the function of many open areas in the pre-industrial settlements were now replaced by those that catered exclusively to amenity and recreation. Parks originated in the late seventeenth century as private residential squares at a time when some cities in Britain were becoming attractive places to live for the upper classes. Among them were the famous Bloomsbury garden squares of London (1775–1850), and the crescents of Bath, developed by the Brothers Wood (1730–67).[7] The development of the public parks in the expanding cities of Europe and the United States in the nineteenth century evolved out of the Romantic movement. They were created in the conviction that nature should be brought to the city to improve the health of the people, by providing space for exercise and relaxation. It was felt that the opportunity to contemplate nature would improve moral standards. A new preoccupation with the aesthetics of natural landscape led to the notion that parks would improve the appearance of cities.[8] The introduction of the Royal Parks in London, Olmsted's Central Park in New York, the Boston Commons and Mount Royal in Montreal are testament to a period of extraordinary social convictions and purpose.

But the continuing expansion of the city since the nineteenth century and the decline of park priorities have created new conditions. There is greater wealth, leisure and mobility among more people, and a desire to escape the city and renew contact with rural settings that the urban park is now unable to satisfy. Work and play have come to be perceived as separate and distinct activities, thus turning recreation into an all-consuming urban occupation. The perception that the countryside exists solely as an urban playground is borne out every weekend when its lakes, forests and farmlands are invaded by people who have little or no direct contact with the landscape as a place of work, or as natural environment. Thus recreation contributes little to the land on which it occurs. Its social effects create conflict between those who earn their living by the land and those who use it for leisure. Its environmental effects are often destructive of streams, soils and vegetation. At the same time the recreational needs of urban people are changing. The urban poor and ethnic groups without access to the countryside, a preoccupation with physical fitness and diversified recreational interests are changing conventional views of how the city's open spaces are used. Recreation, once confined to parks, is now including the entire city.

The city and countryside are mutually exclusive places

'[T]oday it is nature beleaguered in the country, too scarce in the city which has become precious.'[9] It is not hard to understand how this mental dissociation takes place. Perceptually we miss the obvious evidence of natural surroundings, the woods, streams, marshes and fields. We fail, however, to see nature as an integrated connecting system that operates in one way or another regardless of locality, whether this be in the country beyond, or within the city itself.

Water supply and disposal systems leave no indication that the water supplied to the kitchen tap had its origins in the forests and landscapes of upper watersheds, or that rain falling on rooftops and paved surfaces and disappearing without trace into catchbasins and underground sewers is part of a continuous hydrological cycle. The grass and specimen trees of the urban parks and gardens, their plants brought from Korea and the Himalayas, their turf maintained like a billiard table, are difficult to associate with the diverse community of plants that convert sunlight into energy, store carbon and produce the food and materials necessary for survival. The frozen and hermetically sealed plastic package one finds in the meat section of the local supermarket bears not the slightest resemblance to the animal from which it came. Park maintenance usurps natural succession. The regulated air-conditioned climate and tropical plantings of the shopping centre have replaced the cycle of the seasons. Sanitary sewers and the garbage truck break the life cycle of nutrient and materials of natural systems. The urban environment serves to isolate us from an awareness of the natural and human processes that support life.

Yet, the essential creativity of nature, the processes that continue modified and often degraded, continue to function. Rich and diverse natural habitats, remnants from a pre-urban era, occur on quasi-public or private land where public access is restricted and where the gang mower has not penetrated. New communities of plants, species often alien to the region, establish themselves, flourishing in the warmer climate that northern cities afford. These plants, like the mosses, common dandelion, plantain, Tree of Heaven, buddleia, are, in some parts of the world, what we know as weeds. Weeds, botanically speaking, are plants that colonize disturbed land. From a cultural viewpoint, they are plants growing where they are not wanted. They represent the fortuitous communities of the urban environment. Hydrological systems are in evidence in the rainfall impounded on the poorly drained parking lot and playing-field, where the processes of evaporation and groundwater recharge continue the cycle, by accident rather than intent. The sewage lagoon perpetuates the process of decay of organic material and release of nutrients and provides new urban marshland for large populations of shore birds. Flat-topped roofs provide nesting places for night hawks, garbage dumps and waste places attract small rodents which in turn attract the hawks and owls which feed on them. It is these natural systems operating within the city that are the basis of an ecological framework for urban design.

An abundance of energy

The availability of cheap energy has been an overriding determinant of urban form. The energy flow through a city, with its factories, automobiles, heating and cooling systems and high power consumption, is about a hundred times greater than the energy flow through a natural ecosystem.[10] Thus cities place enormous stresses on natural systems, depending on them for resource inputs and for the disposal of unwanted products. There is input of food from agricultural regions and output into the environment of heat and concentrated nutrient energy from sewage treatment plants. Industrial processes draw water from rivers and streams for cooling and return waste heat energy. Solid waste and organic refuse are disposed of in landfill sites using up and often contaminating land and water, and generating large quantities of methane and other gases in the process of decomposition.

With respect to building design, cheap and abundant energy has permitted building to evolve, whose climate, form and style are no longer determined by natural constraints. Steadman has pointed out that the external containing envelope of buildings becomes the consequence of their internal organization and material structure, in contrast to traditional architecture, where the design of the exterior shell was a response to the problem of keeping out the weather and protecting the interior from heat and cold.[11] What was once known as the international style of architecture, which evolved through the decades into other styles, can still be seen in the 1990s in the designed urban landscape. The problem of establishing natural elements in the hostile urban environment of the city centre has produced a landscape whose creation, maintenance and survival depends not on natural determinants, but on technology and high energy inputs. The international style of landscape design that one finds in the urbanizing city edge has little to do with the inherent characteristics of the place that was once there. It is established and maintained in isolation; a predetermined design imposed on its site.

Finally, perhaps the most striking aspect of the city is the amount of wasted energy and effort that is expended to create and maintain such an unrewarding environment. The wealth of opportunity that exists to create a better one is, in the 1990s, only beginning to be explored. The integration of urbanism and ecology achieved through the design and planning process is our concern here. It establishes links between a local and a larger bio-regional view, and makes connections between disparate elements to reveal possibilities that may not otherwise be apparent. The insights that urban ecology provides, when put together with social and economic objectives, creates a rational basis

a

Plate 1.4 *Expressions of the energy urban landscape*

a *The universal turf groundcover and scattered buildings*

b *Loss of visual connections to the countryside*

(Photos: Steve Frost)

b

for shaping the city's landscape. We should now review the principles that seem the most applicable to this view, since they form a frame of reference for the discussion of city form in the chapters to come.

SOME DESIGN PRINCIPLES

Process

Processes are dynamic. The patterns of the landscape are the consequence of the forces that give rise to them: geological uplift and erosion of mountains, the hydrological cycle and forces of water that shape the land, the diversity of plants, animals and people working on the land. The form of the place reveals its natural and human history and the continuing cycle of natural processes. An incident that occurred in the 1960s demonstrates the difficulty that is often experienced in understanding the dependence of form on process.

In 1964, the American Falls International Board convened a conference on the most magnificent urban geological feature known – the falls at Niagara. The issue facing the delegates was the talus that had broken off on the American Falls, forming enormous piles of rock at its base. This event had created much adverse publicity in the press and consequently greatly concerned the branch of the International Joint Commission responsible for its care. The aesthetic of the falls, badly marred in the public mind, threatened to create an international incident. What was to be done? Should the rock be removed? Should it be removed entirely or only partially? And what form should the rocks that remained take to create the most pleasing effect? This and other problems prompted detailed landscape design studies to be undertaken with the help of a large-scale model. The aesthetic impacts of various alternatives were tested over a considerable length of time. The real issue that eventually surfaced, however, was how one perceives natural phenomena in terms of process. It was apparent that the Niagara Falls are a natural feature of astounding drama and grandeur and a source of wonder for all who come to see them. Consciously or unconsciously, this sense of wonder is rooted in the idea that we are face to face with overpowering natural forces, a part of a continuum of geological time and process that has evolved over tens of thousands of years. The erosion processes that originally created the falls continue to do so. What we see today is different from what Father Hennepin saw (the first white man to visit them in 1679), or what our distant descendants will see a hundred or a thousand years hence.

Thus our current appreciation must be seen in this context; a mere instant of time within the evolving continuum of nature. The aesthetic of the falls is a consequence of this evolution. The tendency to view natural phenomena as static events, frozen in time,

Plate 1.5 *The American Falls, Niagara*

The dependence of evolving natural processes on physical form is often not understood. A rock fall in the 1960s, perceived to mar the aesthetic of the falls, caused much adverse publicity and public pressure to 'improve' their appearance

is a root cause of the aesthetic dilemmas that we face. When nature is seen as a continuum, the argument of what is beautiful or what is less so in the landscape becomes, if not meaningless, then of a very different order of meaning. The rock falls at Niagara are simply visible evidence of nature at work. They cannot be regarded as some gigantic engineering toy that has somehow gone wrong and must, within the limits of technological wizardry and irrelevant aesthetic standards, be put right.

The same analogy applies to cities. Urban form is the consequence of a constant evolutionary process fuelled by economic, political, demographic and social change; of new buildings replacing old and old buildings being adapted to new uses, of shifting and changing neighbourhoods, of urban decay and renewal. The concept of process also has radical implications for the city's landscape. The creation and upkeep of unbuilt places have traditionally been seen as a static endeavour; once created the object is to maintain the status quo.

The dynamics of plant communities follow quite different laws that change and evolve in response to natural forces. Thus design and maintenance, based on the concept of process, become an integrated and continuing management function, rather than separate and distinct activities, guiding the development of the human-made landscape over time.

There is a prevailing view among conservation-minded people that human influences on the land are inherently destructive. The disappearance of forests, wetlands and life forms both at home and around the world, leaves one in no doubt that this opinion is, in large measure, well founded. The blunt statement of the manager for state parks in

New York that open space is like virginity – once lost it can never be regained[12] – rings true when we are faced with the destruction of priceless landscapes and cultural heritage in the face of urban development. The preservation and protection of nature can be argued for, in the context of today's values, on the basis of moral and aesthetic values, of the absolute necessity of maintaining genetic diversity and of keeping options open for the future. Design, however, is also directly concerned with the notion of change, and the constructive opportunities that change provides. David Loenthal makes the point that the manager for New York State parks may not be altogether correct in his assertion, since only non-virgins can produce more virgins.[13] This remark contains an important truth when humankind is seen as part of the natural process. Landscapes may be created that are different from the original, but may result, none the less, in diverse and healthy environments. Human beings, as agents of change, have historically been concerned with modifying the land for survival – draining land to create productive fields, exploiting the earth for fuel and raw materials – but often unconscious of the effects of their activities on the original landscape. While the world today exhibits countless examples of destructive change, it is important to remember that there are also many that have been environmentally beneficial. W.H. Hoskins has shown that the origin of the Norfolk Broads, a landscape of water and marshlands in south-east England of great diversity and beauty, was for many years a subject of speculation, one theory being that they had resulted from a marine transgression in fairly recent times. In the 1950s, it was conclusively shown that they were the result of deep peat-cutting in medieval times, some four hundred years ago. Since the region was naturally treeless, peat was a valuable fuel. Water seepage into excavated areas eventually caused the abandonment of peat cutting. Marshes developed and finally enough water filtered in to create the 'artificial' lakes that form the present landscape. In an urban context, one may find flourishing natural landscapes that have evolved from old quarries abandoned long ago. The restored landscapes of the industrial Midlands being brought back into productive use in Britain are examples of the purposeful modification of natural process to bring formally ravaged places back to health. Human or natural processes are constantly at work modifying the land. The nature of design is one of initiating purposeful and beneficial change, with ecology and people as its indispensable foundation.

Economy of means

From an ecological perspective this could be called the principle of least effort. The greatest or the most significant results that spring from an undertaking usually come from the least amount of effort and energy expended rather than the most. It involves the idea that from minimum resources and energy, maximum environmental, economic and social benefits are available. It also involves the idea of doing things small, since it suggests that making small mistakes is infinitely preferable to making very large ones. Over time small mistakes can be adapted to social and environmental conditions; large ones may last indefinitely. Jane Jacobs once predicted that the future city would play the role of supplier as well as consumer of materials, a prediction that is being realized as recycling of once unwanted products has become the order of the day in most cities. Leaves and unwanted organic products are transformed into compost, paper, metals, plastics and glass are put to new uses and products, the city's wasted heat energy is reused to heat buildings and stormwater provides habitat for regenerating landscapes. They become useful resources at less environmental and economic cost than conventional approaches when the right linkages are established.

In developing countries, where poverty and necessity play a large role in the way many people earn a living, economy of means becomes critical to survival. Scavenger groups, for instance, collect and sell for recycling much of the garbage that is thrown away in the cities. There is a good market for recyclable materials such as paper, plastics, bottles and metals. Garbage pickers take their finds to nearby junk dealers who buy their materials and sell them to a variety of industries. For example, used cans are in high demand by kerosene stove-makers. Soy ketchup bottles can be sold back to the factories for refilling. Furniture shops are grateful for new supplies of crates since it is cheaper to use parts of crates than buy new materials. Thus, the recycling process demonstrates the principle of economy of means. It reduces the amount of garbage that must be disposed of in landfill sites, it provides a whole sector of the population with a living, and the process provides environmental, social and energy-saving benefits. The *Jakarta Post* summarized these facts in its report on recycling, 'We have finally arrived at the finding that garbage is decomposable while those components that are not can be used as a resource.'[14]

In an agricultural context, Dutch fruit-growers maintain the grass that grows under their orchards with sheep. The animals keep the grass mown which inhibits competition for nutrients and moisture, and the farmer benefits by having two sources of income. The cutting of grass verges along highway rights of way in the Canadian prairies is another case in point. Agreements between farmers and Highway Departments permit the verges to be maintained at no cost to the public purse while at the same time providing additional hay crops for the farmers. Similarly, a policy in Britain of reforesting the verges of the country's motorways has reduced grass-cutting costs while creating wildlife corridors. This has been found to have benefits in increasing the diversity and movement of plants and animals, and an altogether more interesting visual character at less cost in money and energy.

Plate 1.6 *Diversity of city places*

a *(opposite) Places for wildlife, solitude and education*

b *Places for people, activity and social contact*

Diversity

If health can be described as the ability to withstand stress, then diversity from an ecological perspective also implies health. Odum has commented: 'the most pleasant and certainly the safest landscape to live in, is one containing a variety of crops, forests, lakes, streams, roadsides, marshes, seashores and waste places – in other words, a mixture of communities of different ecological ages'.[15] Diversity makes social as well as biological sense in the urban setting since the requirements of an infinitely diverse urban society implies choice. The quality of life implies, among other things, being able to choose between one place and another, between one lifestyle and another. It implies interest, pleasure, stimulated senses and varied landscapes. The city that has places for foxes and owls, natural woodlands, trout lilies, marshes and fields and urban wilderness, is more interesting and pleasant to live in than one that does not have these places. The city also needs hard urban places, busy plazas and markets, noisy as well as quiet places, playing-fields and formal gardens. It implies that the greater the reliance on a single source of energy the more vulnerable will the urban community potentially become in times of need.

Connectedness

Barry Commoner's well-known principle that 'everything is connected to everything else,'[16] has become, in the 1990s, the embodiment of a larger regional and global view as well as a local one. In the late 1980s the Royal Commission on the Future of the Toronto Waterfront (a commission set up to examine issues along Toronto's lakeshore) recognized the implications of Commoner's principle when it realized that the waterfront could not be viewed as simply a narrow band along the Lake Ontario shore. It is linked by Lake Ontario to the other Great Lakes, by rivers and creeks to the watersheds and by water mains, storm and sanitary sewers and roads to homes and businesses throughout the Metropolitan area.[17] What goes down the sewer in the local residential area has an effect on the watershed, its rivers and lakes many hundreds of kilometres away. The air is influenced by local and regional sources. Beaches, wetlands, cliffs, woodland and meadow along the waterfront are habitat for both resident and migratory wildlife and linked to the hinterland via the river valleys. Human uses of the land – transportation, housing, industry, business and recreation – tie the waterfront to the larger region. To understand a local place, therefore, requires an understanding of its larger context – the watershed and bio-region in which it lies.

Environmental education begins at home

Environmental literacy strikes at the heart of urban life and consequently the way we think about and shape our cities. The perception of the city as separated from the natural processes that support life has long been a central problem in environmental thinking. The urban experience of 'nature' is to a large extent a 'disneyfied' experience: too often relegated to the visit to the zoo, where elephants and tigers, safely behind bars, are on display; too often associated with domesticated pets – poodles, tabby cats, rose gardens and floral clocks. It has been said that children know more about nature in distant lands than they do about the natural things in their own backyards, neighbourhoods and cities. The media reinforces this perception in their treatment of nature. The threatened tropical forests in Brazil, the massive hydroelectric projects that flood thousands of hectares of the Canadian north, the disappearance of countless species of animals and plants in all parts of the globe, remain for many, out there somewhere, beyond the cities, remote from the immediate concerns of ordinary people pursuing the often precarious business of day-to-day living. Environmental education is more than the biology lesson in the classroom, or the yearly trip to the nature centre. It provides no substitute for constant and direct experience assimilated through daily exposure to, and interaction with, the places one lives in. It may also be said that literacy about how the world works is inhibited by the way we have been taught to think about the environment around us and our relationship to it.

This was clearly revealed some years ago when some university colleagues and I conducted a workshop for junior school science teachers on the advantages of environmental education in the city rather than in the 'unspoiled' landscapes beyond the urban areas. The discussion focused on the places close to school where children could learn about nature – the nooks and crannies and abandoned places where naturalized plants and animal communities could be found. At one point, a clearly sceptical teacher asked 'what if there are no natural places near the school, how then does one teach the kids about nature?' One of my colleagues immediately replied, 'Stand them in the middle of the asphalt schoolyard and ask them why they are alive; that would be a good beginning.'[18] The teacher's question highlights a basic problem about how most people think about natural processes. His underlying assumption was that nature is an externality, set apart from human affairs, which can only be studied in non-urban surroundings. Environmental education in his mind had little to do with the interdependence of life systems that includes both human and non-human nature. The novelist John Fowles summarizes the problem in his article 'Seeing Nature Whole'.[19] On a visit to Linnaeus' garden in Uppsala, Sweden, he had this to say about its owner, 'the great indexer of nature' who, between 1730 and 1760, classified much of earth's animate beings:

> Perhaps nothing is more moving at Uppsala than the actual smallness and ordered simplicity of that garden and the immense consequences that sprung from it in terms of the way we see and think about the external world ... for all its air of gentle peace, it is closer to a nuclear explosion, whose radiations and mutations inside the human brain were incalculable and continue to be so.

Fowles suggests that humankind has evolved into an isolated creature, that sees the world anthropocentrically.

> [The] power of detaching an object from its surroundings and making us concentrate on it is an implicit criterion in all our judgments. A great deal of science is devoted to ... providing labels, explaining specific mechanisms and ecologies – in short, to sorting and tidying what seem in the mass indistinguishable one from another.[20]

Seeing nature whole, understanding interrelationships and connections between human and non-human life must, therefore, begin with the places where most people live. The urban allotment garden, for instance, through the daily process of food growing,

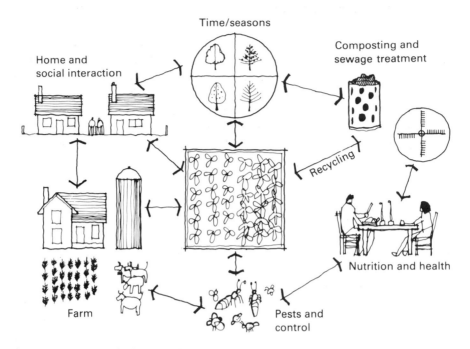

Figure 1.1 *Environmental education begins at home*

The everyday learning experience in the allotment garden that also makes connections to larger regional and international issues

provides a realistic basis for understanding the cycle of the seasons, soil fertility, nutrition and health, the problem of pests and appropriate methods of control. Questions of soil fertility are connected with composting, the source of nutrients in the treatment plant and the recycling of organic materials. The close proximity of food production to home is connected to the energy costs of food production on the farm. The human energy and time invested in urban farming provide economic rewards and social benefits, as leisure time is channelled into productive endeavours. One of the fundamental tasks of reshaping the city is to focus on the human experience of one's home places; to recognize the existence and the latent potential of natural, social and cultural environments to enrich urban places. This provides the best chance of spiritual growth and creative learning since it lies at the heart of environmental education.

Human development and environmental enhancement

There is a common tendency to regard environmentally sensitive design as that process which minimizes destruction to physical and life systems. This idea is also reflected in what we have come to know as pollution indexes. The questions normally posed, for instance, suggest an acceptance of negative values. To what extent can an area be urbanized while minimizing unacceptable water pollution or soils erosion? What are the acceptable levels of contaminants for foods, water or air quality? They imply that some loss, wastage or disruption to the environment is inevitable. It can, of course, be argued that such questions are pragmatic, and based on the realities of current urban conditions since cities are, after all, imperfect, not utopian, places. They may also be useful tools for constraint mapping where the least number of environmental constraints against a proposed use provide a guide to understanding the limitation of a site, or environmental condition. These ways of thinking, however, involve aspects of negative constraint, and inhibit the creative solutions that come from a fully integrated marriage of ecology and human development. Design thinking must go further and ask: how can human development processes *contribute* to the environments they change? Habitat building – creating those conditions that permit a species to survive and flourish – is a basic motivation of all life forms. In nature the by-products of these activities create situations where the altered environment provides opportunities for other species to profit by the change. The action of beavers damming streams, making ponds and cutting forest clearings, has extensive impacts on the forest ecosystem. The temperature of the pond may rise above the tolerable level for brook trout, or the dam may impede the migration of fish up-stream. On the other hand, drowned trees, while they may cause the end of a food supply for some species, create favourable conditions for others. The eventual wet meadow encourages the growth of aquatic plants necessary to support moose. Over time a new succession of vegetation will invade and cover the area. The by-products of one form of life become useful material for others.

In human terms, the negative consequences of human-made change on the environment occur when the necessary linkages are not made. A house or an entire suburban development is an imposition on the land when the resources necessary to sustain it are funnelled through a one-way system: water supply–bathroom tap–drain–sewer system–river–lake or ocean; or food–supermarket–kitchen–dining room–dump. The by-products of use serve no useful functions. The concept behind integrated life-support systems is to make these linkages. They actively seek ways in which human development can make a positive contribution to the environment it changes.

The principles of energy and nutrient flows, common to all ecosystems, are applied to the design of the human environment. The unwanted products of the life cycle become the requirements for another. The recycling of organic products restores soil fertility and

Plate 1.7   *Making visible the processes that sustain life*

a   *Design that conceals natural processes leads to sensory impoverishment*

(Photo: Steve Frost)

b   *Native plant communities created at the Earth Sciences building, University of Toronto for teaching, enriching experiential and scientific learning and environmental awareness*

(Photo: Steve Frost)

its capacity for production. Recycling of used water maintains groundwater levels and water purity. The sewage treatment facility, constructed to collect and process sewage and operating as a human-made wetland, provides a new and rich habitat for a wide variety of wading birds that may not have inhabited the area previously. Stormwater conservation improves the quality of water entering streams and rivers, maintains aquatic life and soil stability and provides opportunities for restoring impaired habitats. Where change can be seen as a positive force to enhance an environment that has been degraded, rather than simply minimizing its further impact or loss, the chances for a constructive basis for urban design will be enhanced. Thus, this principle suggests that development should sustain and reuse the resources it draws on as a benefit, rather than imposing them on the larger environment as a costly liability. Natural processes become internalized into human activities, rather than being incorporated into human affairs when society thinks it can afford it.

The principle of environmental enhancement is also the basis for ecological restoration – bringing natural systems back to a state of ecological health and re-establishing bio-diversity and resilience. Bio-diversity is also linked to cultural history and with restoring both human and non-human habitats in larger bio-regional contexts.

c *Waterfront redevelopment in Capetown, South Africa. Shipbuilding and repairs in a recreational environment making visible the industrial and trading functions of this waterfront*

Thus it is not usually a return to a purely 'natural state', in the absence of human history. It involves the creation of new landscapes – a mix of the natural and the human that may not have existed before, but that recognize the interdependence of people and nature in the ecological, economic and social realities of the city.

Making visible the processes that sustain life

Much of our daily existence is spent in surroundings designed to conceal the processes that sustain life and which contribute, possibly more than any other factor, to the acute sensory impoverishment of our living environment. The curb and catchbasin that make rainwater disappear without trace below ground, cut the visible links between the natural water cycle, the storm sewers that dispose of it into streams and the lakes and rivers that ultimately receive it. We are unaware of the ecological degradation that occurs to aquatic life and to the beaches that have to be closed after a heavy rain. Soft fruits grown in warmer climates and transported thousands of kilometres to cold ones are available in the supermarkets in winter. Materialism (and the ignorance of one's home place that goes with it), may be said to be the product of eating fruit out of season. In Chapter 3 I discuss the significance of community efforts to restore the land, and how the urban forest parks in Zurich are managed to produce saleable forest products in full view of the public. They illustrate the importance of making the natural and human processes that sustain urban life visible. Professor Tjeerd Deelstra, of the International Institute for the Urban Environment at Delft, has suggested that in industrialized countries the management of urban resources is anonymous. The supply of electricity and water, the processing of waste, are not visible to urban people and they consequently do not feel responsible for them. 'You simply turn on the light or open the tap and light and water are there. You buy your food in a shop . . . and put your garbage outdoors and it will be removed.'[21] In developing countries the supply systems – biomass production, brick-making, recycling industries – are all visible. In contrast to the developed countries, the developing world needs basic infrastructures that can integrate the many existing self-help activities and create a self-sustaining urban system. While in wealthy countries the policy of 'bottom-up' public participation must have priority, the opposite may be true in the developing ones – in effect a 'top-down' approach that reinforces and integrates grass-roots initiatives.[22]

Deelstra also makes the point that visibility is essential in economic and political terms. Paying cash to use a toll road is better that paying an annual tax through the bank for the use of one's car. Policies should capitalize on the visibility of the environmental consequences of human actions in the process of daily living. One Dutch city, for instance, calculated how many square kilometres of forest should be planted around it to store the carbon dioxide generated by local household emissions. It showed that there was insufficient available open space in the city to compensate, a fact that reinforced the need to conserve fuel, gas and electricity at home.[23] Thus it may be said that making processes visible is an essential component of environmental awareness and a necessary basis for environmental action.

A BASIS FOR AN ALTERNATIVE DESIGN STRATEGY

In the preceding pages we have seen how, in the presence of cheap energy, the urban environment has been shaped by a technology whose goals are strictly economic rather than social or environmental. This has contributed to an alienation of city and country and a misuse of urban and rural resources. We find a preoccupation with leisure as the prime function of urban parks, while other functions that the unbuilt environment of the city as a whole must serve to maintain environmental quality are largely ignored. Health has been understood to mean the promotion of healthy bodies, not healthy life systems as a whole. We find a preoccupation with aesthetic design conventions that are more concerned with 'pedigreed' landscapes than with the forms that have evolved from the necessity of conservation. The amounts of energy and effort spent creating them does not justify the results when alternatives exist that are cheaper, more effective and more rewarding. Our primary concern is how the city can be made environmentally and socially healthier; how it can become a civilizing place in which to live. As ecology has now become the indispensable basis for environmental planning in the larger, regional landscape, so an understanding of the altered but still functioning natural processes within cities becomes central to urban design. The conventions and rules of aesthetic values have validity only when placed in context with underlying bio-physical determinants. Design principles, responsive to urban ecology and applied to the opportunities the city provides through its inherent resources, form the basis for an alternative design language. They include the concepts of process and change, economy of means that derives the most benefit from the least effort and energy, diversity as the basis for environmental and social health, connectedness that recognizes the interdependence of human and non-human life, making visible the processes that sustain life, an environmental literacy that begins at home and forms the basis for a wider understanding of ecological issues world-wide, and a goal that stresses an enhancement of environmental values that are connected to change – an integration of human with natural processes at a fundamental level.

We seek a design language whose inspiration derives from making the most of available opportunities; one that re-establishes the concept of multi-functional, productive and working landscapes that integrate ecology, people and economy. As environmental issues gain an increasing sense of urgency for the future of cities and for the planet, it is becoming increasingly necessary to meet new goals in the way we shape the future landscapes. Urban land as a whole will be required to assume environmental, productive and social roles in the design of cities, far outweighing traditional park functions and civic values. Many of the problems generated by the city, and imposed on the larger natural environment, will have to be resolved within it. All the city's environmental and spatial elements may then be drawn into an integrated framework, to serve according to their capabilities, as producers of food and energy, moderators of micro-climate, conservers of water, plants and animals, and amenity and recreation. The following chapters will examine various opportunities for achieving this strategy in accordance with the principles that have been outlined here.

2

WATER

•

Wherever you walk in Venice, not far beneath your overheated feet is one of over 22 million wooden stakes, the majority of which are as sound as the day they were driven into the soft silts of the lagoon. The reason for their long lasting service is that the lagoon muds are, and presumably always have been, deficient in oxygen . . . tourists and local inhabitants should rejoice in the fact that the waters of Venice are polluted. All the time there is excess organic matter pouring through the canals, the water will be bung full of bacteria thriving on the products of decay and in so doing using up the oxygen dissolved in the water and that helps protect the all important piles.[1]

INTRODUCTION

It may seem to be a contradiction in terms to be suggesting, as the botanist David Bellamy does, that 'pollution is a good thing'. Yet in many ways the knee-jerk reaction to pollution inhibits a creative approach to the problem. Conventional wisdom needs constantly to be challenged. While it is ironic that Venice should still be standing thanks to its polluted water, this fact reflects a need to explore the physical and biological properties of water, the way its natural cycles are affected by the city and the implications for urban design when alternatives to its current uses and management are examined.

NATURAL PROCESSES

Hydrological cycles

The vast, never ending cycle of distillation and circulation known as the hydrological cycle is a well-known phenomenon. The most important feature is its dynamic quality. Water is constantly being replenished. Evaporating off the oceans it circulates over land masses, falls as rain or snow, percolates below the surface and is returned to the ocean via rivers and lakes. At every point in this movement some water is constantly being returned to the atmosphere as water vapour, to circulate round the earth and fall again as rain or snow. As a result the atmospheric water content remains practically constant. Annual precipitation on the earth's land surfaces averages about 69 centimetres (or 98,300 cubic kilometres of water).[2] However, distribution varies enormously. There are vast desert areas where rain falls only rarely and areas where annual rainfall is as much as 1,000 centimetres. The amount of water evaporating from oceans is on average 9 per cent more than what falls back to the oceans as rain. This 9 per cent represents the amount of rain falling over land areas and which produces the flow of all the world's rivers.[3] The water that falls over the land as precipitation may follow a number of directions. Some of it is evaporated back to the atmosphere before it reaches the earth, some is intercepted by vegetation and is either evaporated, or transpired back to the atmosphere, some filters into the soil and underground reservoirs, and some runs off to enter streams, rivers, lakes and marshes on its way back to the ocean.

Forests and watersheds

Forests protect watersheds. They stabilize slopes, minimize erosion, reduce sediment inputs to streams and maintain the quality and temperature of the water. In the hilly uplands of a watershed, where water sources originate, forest vegetation greatly influences the movement of water from the atmosphere to the earth and back again. It performs a vital function of maintaining stream flows; reducing peaks and potential flooding, but sustaining flow in dry periods. About 30 per cent of the rain falling on the forest is intercepted by the canopy and evaporates back to the atmosphere.[4] Some of the water that does reach the ground percolates through the soil into streams. The root activity and decaying matter of the forest floor act as a sponge holding and gradually releasing a great deal of water. Winter snows trapped in the forest are also gradually released to the stream and rivers in the spring because the ground beneath the forest is less deeply frozen than in open ground and snow melting takes longer in the shade of trees. Some is returned to the atmosphere by the biological processes of transpiration through the leaves of plants. So the forest has a great effect on the movement of water from the atmosphere to the earth and back to the atmosphere. Together with surface and underground water bodies it performs an important storage role.

Human impacts on watersheds

Measurements of water movement in some watersheds have been made for many years. The Canadian Forestry Service has reported on experiments on the eastern slopes of the

Rocky Mountains showing the effect of logging on stream flow. After logging stream flows may increase by as much as 50 per cent, gradually diminishing as the forest regenerates. In one study, thirty-five years elapsed before the stream was back to its natural flow level.[5] The flow regime, or the timing of flows, also changes. Peaks of maximum flow after logging may increase by as much as 21 per cent and low flow by 90 per cent. Erosion increases due to higher run-off and peak flows remove soils and damage productive land. One study of sedimentation from erosional processes showed a sediment concentration 1,000 times greater in streams running through cultivated lands than in streams from a pine forest. Sediment concentrations were seventeen times the pre-logging rate in another study.[6] Thus forests play a vital role in sustaining both the supply and the quality of water.

Biology and transition[7]

Lakes contain practically all the fresh water in existence and maintain the rivers and streams of most watersheds. An understanding of their biological and evolutionary processes is, therefore, useful to this study of urban ecology.

Water biology

Lakes can be classified on the basis of biological productivity. Productivity is a measure of the quantity of life in all forms supported by an ecosystem. At the base of an aquatic food chain green plants convert the energy of sunlight into food calories. The abundance and rate of growth of the algae depend particularly on the supply of dissolved nutrients which ultimately determines productivity at all higher levels of the food chain. Productivity can also be considered to be a measure of the organic matter produced by a system. Each stage of a food chain uses some of the production of the preceding stage, but much decomposes. In a lake, decaying organic matter falls to the deeper water where oxygen is consumed in the process of decomposition. Thus, increased productivity will lead to increased use, reducing the exchange of oxygen between the surface and bottom layers of the lake. Oxygen depletion may proceed to the point at which there is not enough oxygen for various species of fish to survive.

Oligotrophic lakes are nutrient poor. The supply of plant nutrients to the lake is small in relation to the volume of water to which these are added. Productivity is, therefore, low. Usually oligotrophic lakes are deep, often with a mean depth of more that 15 metres. The paucity of nutrients limits the amount of algal growth. The water is usually very clear, and the algal production that occurs can take place through much of the water column as light can penetrate to a considerable depth. As productivity is low,

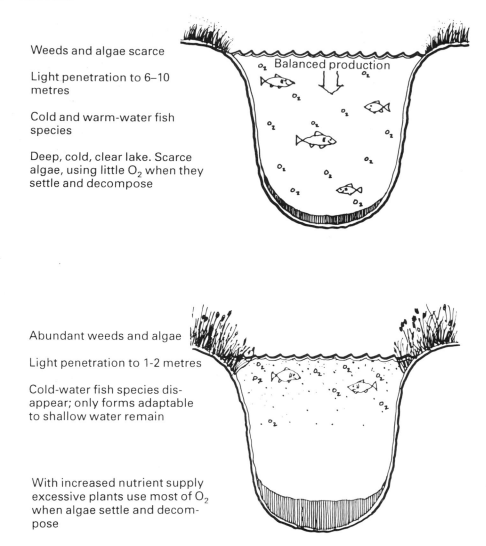

Weeds and algae scarce

Light penetration to 6–10 metres

Cold and warm-water fish species

Deep, cold, clear lake. Scarce algae, using little O_2 when they settle and decompose

a

Abundant weeds and algae

Light penetration to 1-2 metres

Cold-water fish species disappear; only forms adaptable to shallow water remain

With increased nutrient supply excessive plants use most of O_2 when algae settle and decompose

b

Figure 2.1 *Lake productivity*

The differences of biological productivity in oligotrophic and eutrophic lakes

a *Oligotrophic lake*

b *Eutrophic lake*

Source: *Hough Stansbury and Associates, 'Water Quality and Recreational Use of Inland Lakes', prepared for the Ontario Ministry of the Environment, SE Region, May 1977*

decomposing organic matter may not consume a large proportion of the dissolved oxygen in the bottom layers. Fish such as lake trout which have a requirement for oxygen-rich, cool, deep-water habitats, are frequently found in oligotrophic lakes.

Eutrophic lakes are those which are rich in plant nutrients. They are frequently shallow, with a mean depth of less than 10 metres, so they may contain a small volume of water in relation to the supply of nutrients. The abundant supply of nutrients leads to a large algal population, which causes turbidity of the water. As algae require light for photosynthesis, the algal population tends to be concentrated near the water surface, preventing the passage of light to deeper water. In eutrophic lakes scums of algae on the surface of the water are not uncommon, and beds of rooted plants or filamentous algae may cover the bottom of shallow-water areas or bays. High productivity greatly increases oxygen consumption in the deeper water, by decomposition of organic material. In periods of minimal water circulation, either during the summer when the water separates into warm and cold layers (thermal stratification) or during winter ice cover, oxygen concentrations in the lower layers may be severely depleted. This creates conditions in which deep-water fish cannot survive, although warm-water species such as bass, sunfish and pike are found in eutropic lakes.

Lake transition

Lakes are but a temporary feature of the landscape. Even the largest and deepest of lakes are transitory, undergoing a gradual process of change from youth, to maturity, to old age. Progressing even further, the death of a lake can be equated as the onset of swamp or marshland conditions. Thus, the ultimate fate of a lake is to become filled with sediment and eventually be supplanted by grass or forested land. Average natural sedimentation rates for lakes have been estimated at 1 millimetre per year. This means that approximately 10.5 to 15 metres of sediments have accumulated in most lakes since the recession of the last Ice Age some 10,000 to 150,000 years ago. It is not well recognized that changes occurring as a result of natural eutrophication are more complex and subtle and proceed much more slowly than was earlier anticipated. In fact, because many deep lakes, such as Ontario's Lake Superior, have continued· to remain in an oligotrophic state since the last continental glacier receded, natural eutrophication is really an immeasurably slow process. In general, then, healthy bodies of water are self-perpetuating biological communities that purify themselves through the interaction of aquatic plants, fish and micro-organisms.

Human impact on lake transition

In contrast to the slow natural evolution of lakes from an oligotrophic to a eutrophic condition, cultural or human-induced eutrophication can create conditions in decades or less which would take tens of thousands of years in the absence of human activities. The deterioration of Lake Erie, discussed below, is an example of this process. Induced fertilization in thermally stratified lakes of the Canadian Precambrian shield leads to an increased level of phytoplankton, and the onset of high levels of blue-green algae in late summer, and to reduced pH and dissolved oxygen in the deeper waters. Consequently, numerous lakes accumulate growths of planktonic blue-green algae along shore lines that create unpleasant odours when they decompose. Cold-water fish die off owing to reduced oxygen in the lake's deeper waters.

In shallow, naturally eutrophic lakes, increased enrichment resulting from agricultural run-off, urbanization along the system and inadequate sewage treatment increases stresses on already productive environments. When induced enrichment reaches critical values their self-purification capacity is surpassed and rapid deterioration occurs. However, Vallentyne has pointed out that natural and human induced eutrophication differ in two respects – rate and reversibility:

> Natural eutrophication is slow and for all practical purposes, irreversible, under a given set of climatic conditions. It is caused by changes in the form and depth of a basin as it gradually fills with sediment. To reverse natural eutrophication in this sense, one would have to scour out the basin again – a rather formidable task in any man's terms.... Man-made eutrophication, on the other hand, is rapid and reversible. It is caused by an increase in the rate of supply of nutrients to an essentially constant volume of water, without any appreciable change in the depth or form of a basin. As a result, man-made eutrophication can be reversed by eliminating man-made sources of supply. Reversed, however, should not be interpreted to mean anything other than return to what there was before the advent of man.[8]

A classic example of the ability of bodies of water to recover, and how interdependent natural processes and human actions are, was the destruction of the Lake Erie fishery in the Great Lakes Basin. During the 1950s, water pollution due to excessive amounts of nutrients from sewage and farm run-off, caused the disappearance of the mayflies which provide food for many fish and birds. Nutrient enrichment of the water fostered prolific growth of algae and other plants in the lake. This resulted in the breakdown of large quantities of plant matter that used up oxygen and killed other aquatic life. Mayfly predators, such as perch, pickerel and bass, declined dramatically. During the 1970s concerted basin-wide efforts to reduce inputs of phosphorus to the lake gradually improved water quality. As a consequence, the mayflies returned, and the fisheries have recovered.[9]

URBAN PROCESSES

General

It will be clear from the preceding discussion that the biophysical processes of water, land and forests are an interacting system, profoundly influenced by human activity. Since water is a crucial component of the city's support systems, an understanding of these processes is essential to its wise use and management. This is true not only with respect to the larger context of regional watersheds but to the city itself. Many of the pollution problems that affect the water system as a whole begin in the city, so it is here that we must focus our attention. My primary purpose, using this knowledge as a base, is to examine how aquatic processes are altered in cities and what implications for urban design arise from these changes.

There are a number of issues that deserve our attention in a urban context. One is the problem of water supply; another is its disposal. Supply involves the problem of moving water from where it is plentiful – the rivers, lakes and underground reservoirs – to where it is needed in the cities. The natural hydrological cycle is short-circuited by water diversions, artificial storage in reservoirs and urban piped supply systems. Disposal involves the problem of removing it from where it has been used back to the rivers, lakes and oceans via urban drainage systems.

The urban hydrological cycle

Urbanization creates a new hydrological environment. Asphalt and concrete replace the soil, buildings replace trees and the catchbasin and storm sewer replace the streams of the natural watershed. The amount of water run-off is governed by the filtration characteristics of the land and is related to slope, soil type and vegetation. It is directly related to the percentage of impervious surfaces. In forested land, run-off is generally absent, as a glance at the undisturbed litter of the forest floor, even on sloping ground, will show. It has been estimated that run-off from urban areas that are completely paved or roofed might constitute 85 per cent of the precipitation. The other 15 per cent is intercepted by streets, buildings, roofs and walls and other paved and soft surfaces.[10] Piped drainage, designed to carry excess water away from urban surfaces, has two major

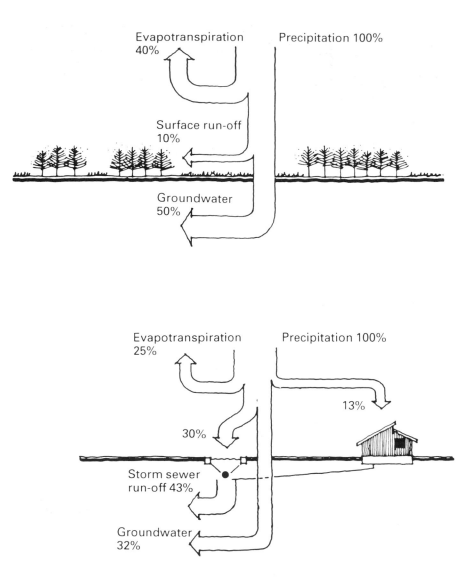

Figure 2.2 *Hydrological changes resulting from urbanization*

a *Pre-urban*

b *Urban*

Source: *Ministry of the Environment,* Evaluation of the Magnitude and Significance of Pollution Loadings from Urban Stormwater Run-off in Ontario, *Research Report no. 81, Ontario, 1978*

Plate 2.1 *Engineered streams as lifeless sewers, designed to reduce flooding in one location compounds the problem downstream*

effects, causing flooding and erosion and impairing water quality, particularly in those climates that suffer from sudden storms.

Flooding and erosion

In urban areas there is a tendency for flash floods and erosion. These are caused by large areas of impervious paving and the concentration of water flows to specific points. The greater the run-off from a storm the more swollen are the streams and the size of flood peaks. Conversely, the greater the volume from run-off, the less there is to replenish groundwater and streams. So, rainfall is accompanied by extremes of flood and low flow. Discharge velocities are also higher than in natural conditions. This is well illustrated in the forested ravine lands that follow a meandering course through densely built-up parts of some cities. In the sudden storms typical of the summer season in Toronto, for instance, valley and ravine streams will rise from a sluggish trickle to a raging torrent in a matter of minutes, carving into already eroded banks and undercutting stone retaining walls. The damaging effects of erosion from water in river courses are greater the more urbanization seals ground surfaces. They increase substantially with the number of urban centres in upstream locations. Downstream solutions to problems caused upstream become more difficult and costly, requiring larger culverts, protection of urbanized floodplains, the straightening and stabilization of stream banks. There is, consequently, wholesale destruction of natural ponds, marshes, plants and wildlife habitats. The effects of urbanization on the water cycle become clear when we see that on average 72.6 centimetres of water per year leaves Ontario cities, of which 31.8 centimetres are from storm run-off. Annual stormwater volumes can exceed sewage flows in low-density urban areas.[11] Calculations show that in an urbanized area that has 50 per cent impervious surfaces and is 50 per cent sewered, the number of stream flows that equal or exceed the capacity of its banks would, over a period of years, be increased nearly fourfold.[12]

Water quality

Conventional storm drainage systems impair water quality and disrupt aquatic life. Water pollution of lakes and streams results from combined sanitary and storm sewer systems that still serve many cities. The provision of overflows permits mixed sewage and stormwater to bypass sewage treatment plants in major storms and enter rivers and streams. The wet-weather loads from combined systems may be many times larger than loads discharged from treatment plants during storms and can equal or exceed total annual discharges from treatment plants.[13] Water quality is also affected drastically by sedimentation. Surface run-off from land cleared of forest increases overland flow and consequently the amount of sediments and nutrients carried to streams from agricultural land uses. The result includes loss of soil, turbidity, accumulation of sediments and higher temperatures in water bodies which makes them highly eutrophic. The rate at which urbanization can change rates of soil loss (in kilos per hectare per year) from previously undeveloped land can be seen in the following list:[14]

forest land	5.5–110
cultivated agricultural land	110–4,360
exposed construction sites	552–92,800
developing urban areas	92–2,208
developed urban areas	32–160

It has been estimated that in the state of New Jersey the sediment yields that can be expected for forested areas in flat land are 10.2 to 41.5 tonnes per 2.6 square kilometres annually, while those in moderately heavily urbanized areas are 25.5 to 101 tonnes.[15] In addition to sediments a wide variety of chemical pollutants, salts, heavy metals and debris add other contaminants to the water. They are carried through storm sewer systems and contribute to higher temperatures, BOD (biological oxygen demand) and a marked depletion of aquatic life in rivers, streams and lakes.

SOME PROBLEMS AND PERCEPTIONS

The development of a reliable water supply has been a primary determinant in the growth of large and densely populated cities. It has provided the means for controlling disease, raising public health standards and effectively fighting fire. At the same time, abundance and security of supply lead to the perception of water as a free commodity, and result in misuse, wastage and environmental pollution – a fact that is particularly true of cities well supplied with water. A review of how water systems in the city have evolved and how they have influenced urban life is, therefore, appropriate here.

Keeping clean

Whole cultures have evolved around the ritual of washing. Superb monuments in engineering, architecture and art have been its product. The best-known historically are, of course, the Roman baths that were built in many of the cities the Romans occupied. As the great engineers of antiquity, they were concerned with hygiene and public health. Indeed, life in large cities such as Rome would have been impossible without a water supply and sewers. Although the first recorded water-supply system was built in 591 BC to water the fields and palace gardens near the city of Nineveh,[16] it was the Romans who recognized the need for clean water and secured their supplies from distant mountain streams via aqueducts, rather than from the River Tiber. The first was built in 312 BC by the censor Appius Claudius. These supplied the baths and city water. The Cloaca Maxima, Rome's famous sewer, was built about 600 BC and emptied into the Tiber. It is interesting to note that from that time no one bathed in the river or drank from it. By 226 AD no less than eleven aqueducts carried water to the city.[17]

In the centuries that followed, towns in the medieval and Renaissance eras were notorious for their lack of sanitary facilities. The streets acted as the dumping ground for the refuse of the town. However, as Mumford has observed, in spite of these practices the smallness of medieval towns, their accessibility to fresh air and open countryside and

Figure 2.3 *The ancient well in Piazza Cavour, San Gimignano, Italy*

Figure 2.4 *The fountains of Versailles, France*

The traditional role of water in cities – for drinking and social focus – became subjugated to its purpose as art, serving the extravagant tastes of the nobility. It is said that when Louis XIV walked in the gardens, each sector of the Versailles waterworks was turned on just before the king arrived. There was never enough water to keep all the fountains running at the same time

the importance of the ritual bath maintained greater health than might normally be expected – a situation that could not be maintained as cities grew larger.[18] Water was supplied via the public fountain which served three purposes: for drinking, as a centre of social life and as a work of art. The tradition of the ritual bath declined in the sixteenth century, which was marked by poor personal hygiene, and was reintroduced into England in the seventeenth century as a luxury. Bathing was seen as a curative rather than a cleansing process.[19] With this in mind, one can surmise that the extensive use of perfumes by royalty in the Court of Versailles (which became known as 'Le Cour Parfumé')[20] undoubtedly had a functional role to play. While important suites in the palace were equipped with bathrooms, these were dismantled in the reign of Louis XV – the Age of Reason.[21] Most of the palace's water supply was directed to the fountains, whose waterworks are said to have developed 100 horsepower with a capability of raising 4.5 million litres of water a day 150 metres.[22] In effect, the traditional role of the fountain as a supplier of water for drinking, and as a centre of social life, was replaced by its sole use as a work of art in the pedigreed landscapes created by nobility.

The private bathroom and the health standard of a WC for every family were made possible by the introduction of sanitary sewers in the nineteenth century. The sewer, introduced into many cities in the eighteenth century, was built to carry away rainwater, so it antedated the development of the sanitary sewer. Until this time human wastes were emptied by 'night soil men' hired for this purpose. The relationship of this practice to farming has been described by Tarr.[23] The waste, collected from households, restaurants and markets, was sold to neighbouring farmers for use on their land. The law often stipulated that cesspools could only be emptied at night, hence the name 'night soil'. The practice was widely followed in the New England and mid-Atlantic states of America, where wastes were collected in seventy-four cities. Baltimore fertilized garden crops with urban night soil as late as 1910. Tarr reports that 70,000 cesspools and privy vaults were emptied and sold to a contractor for 25 cents per load of 900 litres. Virginia and Maryland farmers bought over 54.5 million litres a year to grow crops such as potatoes, cabbage and tomatoes, which were subsequently sold in the Baltimore market.[24]

As cities became more densely populated, the cesspools proved incapable of handling the increased load of human waste. The introduction of piped water vastly increased consumption. The estimated consumption of 13.5 litres a day before piped supplies were introduced rose to between 180 and 270 litres. Also, the consequent health hazards from cholera and yellow fever epidemics that periodically swept the nineteenth-century cities brought about the crusade that forced the cities to build the sewerage systems.[25] When connected to every household, they removed raw wastes directly to the rivers and lakes. Thus, public health in the cities improved. In London prior to 1850, the old tributaries to the Thames were used only for surface drainage. By the middle of the nineteenth century, largely as a result of the widespread use of WCs and a rapid expansion of the population, the River Thames received the untreated sewage of 4 million people. It is reported that in 1858, known as 'the year of the great stink', it became necessary to hang sheets soaked in disinfectant at the windows of the Houses of Parliament to counteract the smell.[26]

The problem of water use in cities has several facets. The first is the continued growth and improvement of technology. Today, consumption of water for domestic purposes is estimated, on average, in North America to be 1,135 litres per person per day. The largest consumer of water is the bathroom at 340–450 litres per day per person. It takes 90 litres to take a bath, 45 litres to do the dishes, 23 litres to flush the WC. The requirement that water must be of drinking quality regardless of its use, increases pressure on the natural system for its continued supply, and on water filtration technologies for its standards of potability. The same quality of water services fire fighting, car washing, irrigation, domestic and industrial uses.

With this kind of use come the immense physical and biological problems associated with the return of used water to the natural system. The sewage treatment plant has provided a partial technological solution to the immediate problem of contaminated urban water. The increasing quantities of pure water taken out of the natural system and returned contaminated have been perceived as a problem requiring engineering, rather than biological solutions. The effect of removing the problem of disposal away from the city, while undoubtedly improving health and eradicating epidemics in western cities, has also delayed solutions to the ultimate, larger problem of wasted resources.

Keeping your shoes dry

The storm sewer and catchbasin have for decades remained the conventional method of solving the problem of urban drainage and water disposal and have, until recently, been unquestioned. As the established dictum of engineering design, the rules have been simple – water drains to the catchbasin. It is here that, perceptually, the problem stops and connections with larger environmental issues of watersheds are not made. In commenting on the fragmentation of environmental concerns, human activities and policies that beset cities, the Royal Commission on the Future of the Toronto Waterfront began its first report this way:[27]

> At five o'clock in the morning in early July, the rain began, slowly at first and then with increasing intensity. It struck rooftops and trickled down gutters, gathered on driveways, parking lots and roads. Along its way the swirling stormwater picked up animal feces and herbicides from parks and yards, as well as asbestos, oil, and grease from roads. Before the rainfall ended, 4.5 billion

litres of rain water had gushed into the labyrinth of storm sewers under the metropolis.

Between seven and nine o'clock, people began to rise, taking showers, brushing teeth, and flushing toilets in 1.5 million households. By eight o'clock, when most had left for work or school, 770 million litres of wastewater had gone down household drains and into the sanitary sewer system. Combined storm and sanitary sewers were overflowing, and a noxious brew of stormwater and untreated sewage was flowing into local rivers or surging towards the sewage treatment plants. By nine o'clock, the hopelessly overburdened treatment plants began to bypass partially treated effluent directly into the nearshore of Lake Ontario.... Unseen by commuters, the brown and swollen rivers in the area disgorged their loads of sediments and toxic chemicals into Lake Ontario. At the river mouths, fishermen tossed their catches back into the lake, mindful of the signs that warned against eating fish. 'Just a reminder to stay out the water at area beaches for two days after this rainfall' the radio voices continued. By mid-morning, public health officials would be testing water at the beaches lining the waterfront; in less than a week, many would be closed to swimmers.

'Cloudy this morning, sunny later with highs of 25 degrees'. Along with the afternoon sunshine would come high levels of eye-stinging smog. 'And cooler temperatures tonight, especially near the lake. All in all,' said the news readers, 'a pretty average day in Greater Toronto.'

It is clear from this commentary that there are serious problems of discontinuity in our perceptions of urban systems and natural processes. The benefits of well-drained streets and civic spaces are paid for by the costs of eroded stream banks, flooding, impaired water quality and the disappearance of aquatic life. The quite understandable human penchant to keep pets, for instance, is one of the many factors responsible for degraded water quality. The products of animal defecation in public spaces, washed untreated into urban storm sewers, create health hazards for people, fish and aquatic life in general. There has been a failure to grasp fully the hidden environmental and economic costs of local water management practice, such as connecting downspouts to the sewer rather than discharging roof water directly to the ground. The annual costs in erosion control, channelization of streams and underground stormwater systems are the engineering consequences of the need to keep one's shoes dry. Conventional urban design, in fact, contributes to the general deterioration of the environment by shifting an urban problem on to the larger environment, and by the failure to recognize and act on the relationships between human actions and natural systems.

There are several ways of approaching these problems that differ from current practice. The first is the obvious and well-established conservation measure of using less, and many cities in the 1990s have adopted water economy policies. Tucson, Arizona, for instance, a city that is entirely dependent on diminishing groundwater reserves, instituted conservation measures in 1980 to reduce municipal, industrial and agricultural water use. To encourage economy, water rates are increased to match water use by residents. Between 1976 and 1986 this resulted in a per capita drop from 932 litres per person per day to 705 litres.[28]

The second way of approaching water problems is related to the perceptions and values that have evolved from urban life. The traditional role of the fountain as a vernacular expression of water supply, social interaction and art became subjugated to an expression of art alone in the pedigreed gardens created by the rich and powerful. The aesthetic use of water has remained separated from its functional uses. The preoccupation with the expression of water as display or status symbol, defying gravity with extraordinary feats of engineering, is reflected in the modern city. The sparkling fountain gracing the civic square, symbolic of pure mountain streams, cascading falls and unspoiled places, is made possible by engineering technology – the filtration plant, hydraulic equipment and heavy doses of chlorine. While accepting the validity of its aesthetic purpose, one may well question some other assumptions and values that it represents.

The dichotomy between the euphoric image of nature and the realities of the urban hydrological cycle emphasizes in another way the isolation of urban life from natural processes. It is difficult to reconcile the image of sparkling fountains and children's play pools with debris-clogged and muddy streams or the blackened snow that piles up along streets in northern cities over the winter. This sense of isolation has also been aggravated by municipal design and practice. Keeping your shoes dry in the city ensures that people remain unaware of where the water comes from or where it goes. Water is drained off streets, parking lots, pavements, plazas, school yards, front and back gardens and parks, and disappears from human consciousness, perpetuating environmentally destructive practices. The way water and other urban life-support systems are used are not apparent: 'You simply open the tap and the water is there.'[29] The principle of visibility discussed in Chapter 1 is, therefore, critical to environmentally responsible behaviour. Thus, the third way of approaching the question of water is to consider the opportunities for urban design that arise when the problems of disposing of the city's 'waste' water become opportunities for restoring hydrological and ecological balance, and enriching the experience and complexity of the city as a place. These alternatives will be examined on pp. 70–80.

SOME ALTERNATIVE VALUES AND OPPORTUNITIES

The Thames revival

The deterioration of rivers from urban pollution is a fact that has been attributed to the costs of progress. But one of the most interesting examples of the process of deterioration and the seemingly impossible feat of bringing a river back to health is the story of the rehabilitation of the River Thames.[30] The 40 kilometres of the tidal river within London are subject to daily fluctuations and have been subjected to two periods of gross pollution: once in the 1850s and again in the 1950s. As an ecological entity the Inner Thames was a wilderness of marshes and reed beds, harbouring vast populations of birds. It is reported that spoonbills nested in the area of Putney Bridge up to the sixteenth century and montague and marsh harriers hunted in the marshes of south London. The water supported a thriving fish industry including salmon and sea trout. By the middle of the nineteenth century the widespread use of WCs and the rapid expansion of the population created such polluted conditions that all fish were eradicated, together with the birds that fed on them. The first effort to build a sewage system was completed in 1874. Its main feature was the construction of intercepting sewers that discharged into the Thames at the extreme east end of London at Beckton and Crossness, avoiding the central areas.

By 1900, some of the river's quality had been restored by this work, but it gradually deteriorated again during the first half of the twentieth century due to increased discharges of effluent from both domestic and industrial sources. During the 1940s and 1950s the health of the Thames was at its low ebb, little better than an open sewer, containing no oxygen and permitting the survival of only specialized forms of life adapted to anaerobic conditions.

The crisis conditions of the 1950s led to the creation of several government committees which surveyed the Thames and examined the effects of pollutants. As a result of these investigations the problems of large pollution loads in an enclosed tidal system were better understood, and a programme of improvements was drawn up. It was recognized that the effect of pollutants on dissolved oxygen in the river was a critical factor. As we saw earlier in this chapter when discussing the induced eutrophication of lakes, when the oxygen content is entirely removed, oxidation of wastes cannot occur and anaerobic conditions are created. In this situation hydrogen sulphide is formed, giving off the familiar smell of rotten eggs. The problem of fluctuating tides exacerbated the problem on the Thames since it may take up to eighty days for water to be flushed to the sea in periods of low rainfall. In 1964 greatly enlarged and improved sewage works were begun, and these were completed in 1974. The most up-to-date filtration, treatment and aeration equipment was installed so that the fluid discharged into the Thames would be pure water.

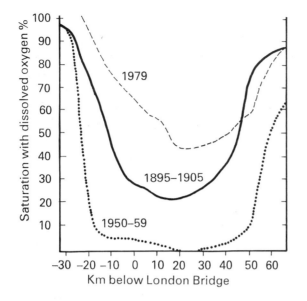

Figure 2.5 *Oxygen sag curves, third quarter*

The improvement in the quality of the River Thames since the last century can be seen in this graph showing the percentage dissolved oxygen for the summer quarters of the three periods 1895–1905, 1950–9 and 1979. The river is now healthier than before records began and represents a return to the quality which was last shown in the mid-eighteenth century

Source: *L.B. Wood, 'Rehabilitation of the Tidal River Thames', unpublished Thames Water Authority Paper, n.d*

Two criteria for estuarine quality became the basis for control. First, quality must be good enough to allow the passage of migratory fish at all stages of the tide. Second, it must support fauna on the mud bottom, essential for sustaining sea fisheries. The concept of 'pollution budgets' was introduced: these set the maximum quantity of pollution load for treatment plants that can be tolerated if the desirable water quality is to be secured. A goal of 30 per cent dissolved oxygen was set to reach the desired quality. A third criteria was that toxic and non-biodegradable substances such as heavy metals should be excluded from industrial effluent. Records of oxygen sag curves for the Thames have shown that the objectives have been met. The curve for the third quarter of 1979 showed that despite low freshwater flows, the average minimum dissolved oxygen was above 44 per cent. The Thames that had been devoid of fish for 48 kilometres between 1920 and 1964 now supports aquatic life. By 1975, no less than eighty-six species of fresh-water and marine fish had been identified. Large flocks of wildfowl returned to the Inner Thames and as a wintering area it was attracting 10,000 birds or more at peak times.

Impressive as the Thames revival was, however, societal values in the 1980s, and early 1990s, began to reflect broader international concerns for the health of city nature. Vallentyne's comment that 'river ecosystems are integrations of everything that takes place in their drainage basins'[31] reflects the reality that restoring rivers to a state of ecological health involves more than resurrecting downtown waterfronts, or reducing pollution, or making drinking-water less poisonous – a single-minded policy objective of many government agencies concerned with water quality. It involves an environmental view that sees rivers as life-sustaining natural systems, where value cannot be measured in terms of short-term economics, or how much electrical energy can be extracted from river diversions. It involves a view that sees economic objectives linked to social and environmental ones, where urban development processes contribute to, rather than detract from, the environments they change, where economy of means dictates the use of water, and where environmental education can begin by protecting life in the valleys and watersheds. The following case-study illustrates how this holistic and ethical view of urban natural systems, combined with a growing determination for political empowerment by local people, can achieve integrated, visionary and pragmatic strategies for the rehabilitation of a river.

The healing of a river: the Don Valley, Toronto[32]

The Don River is one of a system of watersheds in the Greater Toronto Bio-region that extend from the Oak Ridges Moraine (the aquifer recharge area that forms their rural headwaters) to Lake Ontario. The Don is significant to the City of Toronto in that it is the most highly urbanized and degraded river, particularly in its lower reaches, in the Greater Toronto Bio-region. It is a river whose essential natural values have been ignored and despoiled for over two hundred years: its once pristine waters now badly degraded from storm and combined sewers; its lower valley channelized and ransacked by an expressway, a four-lane road, railway tracks, transmission towers and salt dumps; its vegetation and wildlife diversity greatly impaired; its sense of wholeness, beauty and place a fond memory for those who had known it that way. Moves to restore the river became an act of faith by the citizens of Toronto that grew out of the concerns of many people for the natural heritage of their city. Beginning as an informal citizen's organization, 'The Task Force to Bring Back the Don' was formalized and supported by Toronto City Council in 1990. Its purpose was to begin the process of renewal of the most degraded part of the river that flows through the City of Toronto, and ultimately to initiate the restoration of the entire watershed.[33]

The Task Force had three primary goals. First, to re-establish ecological diversity in the lower valley in ways that would integrate its cultural history with human and non-human values; second, to develop recreational and educational strategies that would be in tune with the essential nature and functions of a river valley; third, to reconnect the river to the lake. The Task Force envisaged an estuarine marsh at its mouth as an expression of this goal. These goals required an understanding of the valley as a whole, and the interconnectedness of its parts as a 'natural' and 'human' system. The concept of restoration, therefore, involved taking more dramatic environmental action where a 'return to a natural state', in a literal sense, was not practically feasible, or maybe even

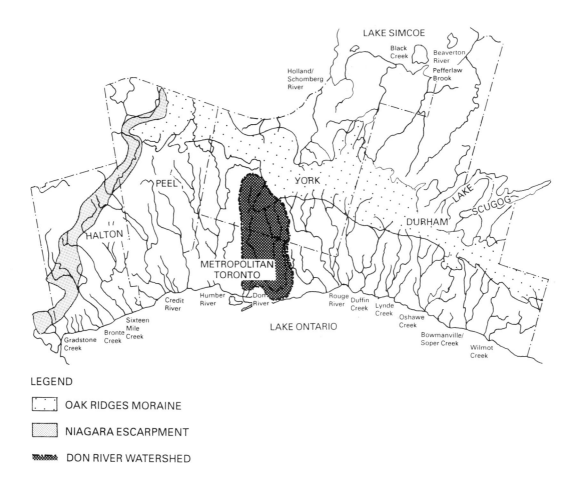

Figure 2.6   *Greater Toronto Bio-region showing the Don River watershed in its larger context*

Source: *Royal Commission on the Future of the Toronto Waterfront*, Watershed, *Interim Report, Toronto, August 1990*

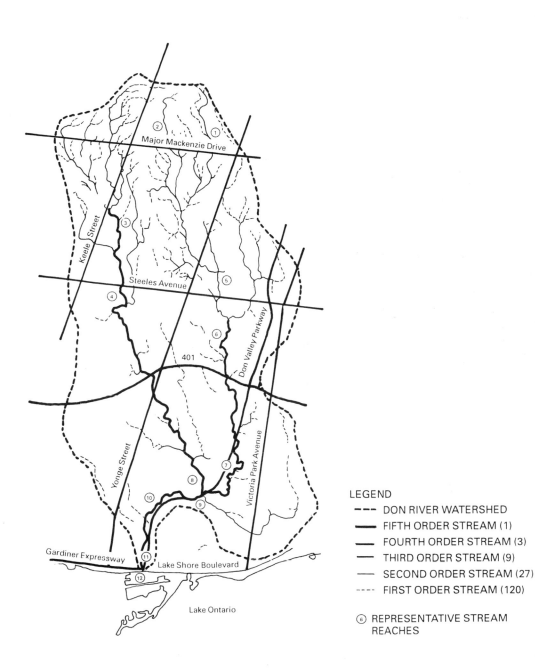

Figure 2.7 *The Don River Watershed*

Source: *Task Force to Bring Back the Don, 'Bringing Back the Don', Toronto: Hough Stansbury Woodland, Prime Consultants, in association with Gore and Storrie Ltd, Dr Robert Newbury, The Kirkland Partnership, 1991*

desirable. It required both major and small-scale interventions, continuing over many years, to return the Don river to a state of health.

Historical background

Anyone walking through the Lower Don today would be hard-pressed to recognize it as a natural valley. But to bring it back to health, it is necessary to understand its past – what the river was and how its present plight came about. We have to peel back the rubber mask of today's altered river landscape[34] – the structural changes that have been imposed over its original form – to reveal its original underlying processes and patterns.

The river's glacial past

At various times, the Toronto area was covered with shallow seas, glaciers hundreds of metres thick, and freshwater lakes and rivers that had basins larger than those of today. Different plants and animals have inhabited the area, responding to changes in climate and land migration routes. Each left its own signature of sedimentary deposits and fossils in the geological record. During the Pleistocene epoch which began a million years ago, three successive waves of glaciation buried the bedrock beneath thick glacial till. The

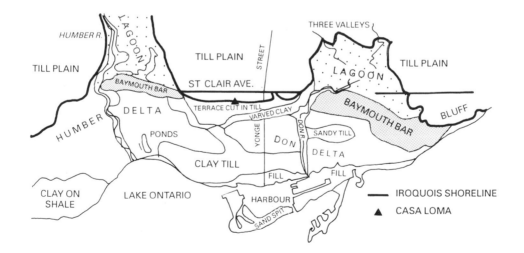

Figure 2.8 *Physiography of the Toronto area*
Note the shore of early Lake Iroquois in relation to the present shoreline
Source: *L.J.Y. Chapman and D.F.Y. Putnam,* The Physiography of Southern Ontario, *Ontario: University of Toronto Press, 1966*

Plate 2.2 *'The Artist's Choice' on the Don River*

The river was once seen by a few as a place of beauty and inspiration (compare with Plate 2.4b, p. 61)

(Source: Postcard, c. 1905)

Don was born at the end of that time, only 13,000 years ago. As the glaciers retreated streams began flowing south from its source, a porous water-filled ridge of glacial debris named the Oak Ridges Moraine. In the river's early days it flowed as two streams into a lagoon formed by a sandy baymouth bar that had been created by wave action and shoreline currents, and thence into the early glacial Lake Iroquois. As the glaciers continued melting, the land began slowly lifting up and Lake Iroquois shrank to become the present Lake Ontario. Now the Don flowed as one river out of its old lagoon and south across the flat sediments of what had been Lake Iroquois. As it entered Lake Ontario as one stream, the process of building a baymouth bar and backshore lagoon was repeated, forming the harbour islands spit and a protected lagoon known as Ashbridges Marsh. This was a vast fertile wetland habitat for fish and wildlife, and an important food source for the early Indian inhabitants of the land.

Early settlers

Toronto was first settled in 1787 when surveyors laid out a city plan for the future capital of Upper Canada, and in 1793 John Graves Simcoe became its first Lieutenant Governor. The settlers harnessed the river's energy, built mills for lumber, flour, wool and paper and

Plate 2.3 *The original pre-development river meanders imposed on the present channel*

(*Photo: Steve Frost*)

mined the valley's shale for brick-making, from which much of the early city was built. In less than 150 years, they cleared the lower valley of merchantable trees. The river was also perceived as a threat and an obstacle. Floods swept away mills and bridges, the river was an obstacle to the eastward expansion of the city and the great wetland, its mouth reviled as an unhealthy swamp, lent credibility to the argument that straightening out the river and filling in the marshes would 'secure the sanitary condition ... to the said river'.[35] By the end of the century, engineers had turned the last 5 km of the river's meanders, where it dropped its sediments, into a canal. The railways were built in the valley, and the Ashbridges Bay marshes were filled in to create the port lands, the most massive engineering project on the continent in its time, forcing the Don into a right-angle turn into the harbour. Today, the city has turned its back to the river; it has become a gap between places, rather than a place in itself. As a sensory experience, it has become

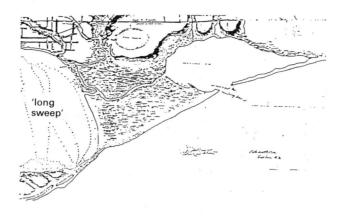

'long
sweep'

Figure 2.9 *Early sketch of Toronto showing the relative positions of its present and proposed defences in 1846 is roughly drawn but purports to show conditions within the Ashbridges Marsh*

'Ashbridges Bay' is named for the first time. The Don in this map has two outlets, that in the north-east corner of the harbour being shown for the first time although it was said to have been opened for tactical purposes during the war of 1812

Source: *T.H. Willans, 'Changes in Marsh Area Along the Canadian Shore of Lake Ontario',* Journal of Great Lakes Research, *vol. 8, no. 3, 1982, p. 573*

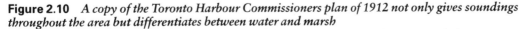

LAKE ONTARIO

Figure 2.10 *A copy of the Toronto Harbour Commissioners plan of 1912 not only gives soundings throughout the area but differentiates between water and marsh*

Cottage settlements on the peninsula and the sand spit are shown as are early industrial enterprises. The north outlet of the Don has been closed and the river diverted south to the Keating channel, making a right-angle turn before entering the inner harbour of Lake Ontario. Work on the channel has straightened the irregular edges of the mainland and landfilling has already consolidated the marsh to the north and west of the river/channel intersection. Plate 2.7 shows the present day industrial area on the original marshes. It also shows how the verdant green of the river upstream contrasts starkly with the denuded character of its last downstream stretch

Source: *T.H. Willans, 'Changes in Marsh Area Along the Canadian Shore of Lake Ontario',* Journal of Great Lakes Research, *vol. 8, no. 3, 1982*

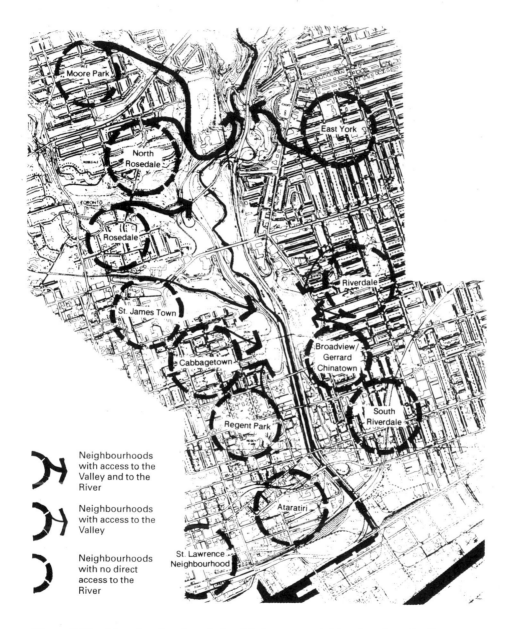

Neighbourhoods
with access to the
Valley and to the
River

Neighbourhoods
with access to the
Valley

Neighbourhoods
with no direct
access to the
River

Figure 2.11 *Map showing the relationship between neighbourhoods and valley access*

It reinforces the Task Force's concerns that the river is a gap between places, not a place in itself. Much of the surrounding development has its back to the river inhibiting access to the valley

a forgotten place; unloved and unused. The roads and expressway, the legacy of the 1950s, have made the valley a transportation corridor, inhibiting access for walkers and cyclists. Other facets of present-day conditions provide an overall image of the river's environmental conditions.

Water quality issues

The watershed is 70 per cent urbanized. The river and its tributaries are consequently subject to various potential sources of pollution from discharges from some 1,185 storm and combined sewer overflows emanating from municipalities north of the City of Toronto within the Don's 36,000 hectare watershed. Ninety-five per cent of the Don's pollution originates from these sources.[36] In addition, there are flows from an existing water pollution control plant, snow dumps, landfill sites, agricultural and rural sources. Estimates indicate that given an approximate removal efficiency of pollutants of 50 per cent through the elimination of combined sewer overflows and treatment of stormwater, the City of Toronto on its own could expect to achieve a 2.5 per cent reduction in annual loadings to the Don River. This shows that the long-term health of the river is heavily dependent on the co-operation of jurisdictions within the entire watershed (Figure 2.7). Policies to improve water quality and to control peaks and lows of river flow depend, therefore, on mutual co-operation of municipalities throughout the watershed.

Vegetation and habitat

Once forest cover played a vital role in maintaining the health of the drainage basin and the quality of its waters. It aided the infiltration of rainwater into the water-table and was important in controlling the rate of flow of the river. It kept the streams cool and maintained the diverse fish life and yearly salmon runs about which Lady Simcoe and others marvelled in the early days of settlement. Wetlands provided an immense diversity of aquatic flora and fauna, as reservoirs of fresh water and helping to even out the extremes of stream flows after heavy rain.

Most of the native vegetation of sugar maple, white pine, beech and oak in the lower valley has long gone, cut for farmsteads and settlements and export to England. Similarly, the wetlands that were associated with the bottomlands and river delta have also disappeared. Fish and wildlife populations are no more than a mere shadow of what they once were. Yet, a record of the original forest and wetland communities that once dominated the area still exists in the ravine tributaries and places isolated by the expressway in the valley. The valley continues to serve as a migratory corridor for wildlife, and mammals, butterflies and foxes are common in the forested ravines. Spring and autumn birds and monarch butterflies migrate through the corridors, hawks prey on

Plate 2.4 *Perceptions of the Don*

a *The urbanized channel: a forgotten place of industrial buildings, advertising signage and expressway interchanges. Early in Toronto's growth the river was seen as a barrier to eastward expansion of the city. It was straightened and encased in a channel by the end of the nineteenth century to facilitate the development of industrial land and to create another railway entrance to Toronto*

small mammals in open meadows, herons, kingfishers, ducks and mergansers are to be found along the river banks. This life provides an indication that the river, while impaired, is far from dead and is, therefore, a viable candidate for restoration.

The valley as sensory experience

The Task Force spent countless hours in the Don Valley documenting its cultural, industrial and natural features, evidence of its destruction and naturalization and its existence as sensory experience. Among the images documented as planning proceeded were:

- it was a place difficult to get into but, once in, almost impossible to get out of;
- the high levels of noise from the expressway where one had to shout to be heard;
- the magic of places made fortuitously inaccessible by road alignments;
- the presence of industrial objects: fences, dumps, PCB storage, oil drums, switching stations, advertising signs for the benefit of drivers on the expressway;
- the discovery of unusual plants and the sighting of a heron;
- the magnificent engineering of an early bridge crossing the valley;
- the coolness of the valley on a hot summer's day.

b *The floodplain further upstream: the Don reasserts its meanders and the valley begins to look like a river. In the 1950s it became the site and route for the Don Valley Expressway and associated interchanges. The Bloor Viaduct, whose construction was described in Michael Ondaatje's book* In the Skin of a Lion, *is today a noteworthy example of early twentieth-century engineering heritage*

c *Heavily forested upper reaches of the Don begin to develop a sylvan character reminiscent of the natural river*

These impressions, both negative and positive, represent the framework of the history that has made the valley what it is. It was apparent that the opportunities for rehabilitation were fundamentally linked to this legacy.

Land use and urban policies

It was also apparent that the Lower Don had never been perceived as a single connected valley in current planning policies. It was seen as an edge to a wide variety of planning districts, a transportation corridor unprotected by Toronto's Official Plan. It had, therefore, no status as a place in its own right, worthy of its own policies as a river valley. These circumstances, and a complex ownership pattern of public and private lands and often conflicting land uses within and surrounding the lower valley, influenced the long-range strategies for its restoration.

Strategies for the Lower Don

Restoration strategies were based on a number of interrelated courses of action. These were intended to re-establish its health and diversity, and bring the valley and its river back to the city, so that it could again be treasured and experienced as a valued and essential part of urban life. They were based on establishing a process of design strategies and policies that would allow remedial action to continue into the future.

Observation revealed the existence of three clearly defined landscape types or units, each intimately connected to the whole, yet each with a distinctive natural and cultural character of its own:

- the mouth of the Don where it makes a right-angle turn into the lake, and meets the elevated expressway and the portlands;
- the channelized and physically restricted section of the river;
- the upper section where the river maintains its original meanders and the valley broadens out into a major floodplain.

These became the basis for future planning strategies. But first the river's hydrology had to be understod as a whole if the goal of reconnecting the river to the lake was to be realized.

Hydrology concepts

The overall strategy for restoring habitat, open space and a delta/marsh in the lower valley depended on modifying the hydrology of the river using natural principles of

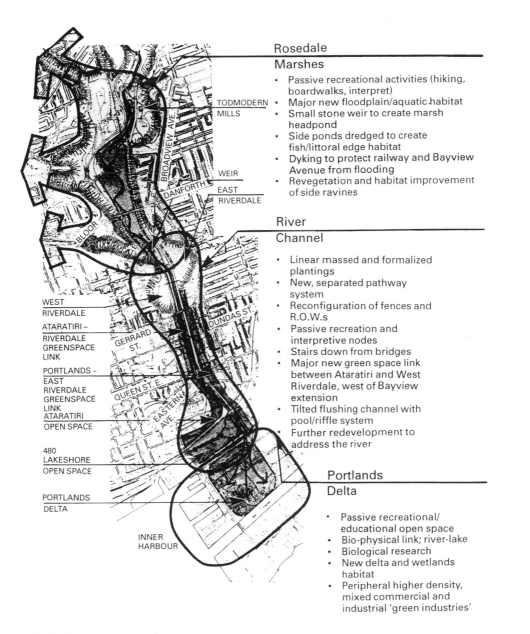

Rosedale

Marshes

- Passive recreational activities (hiking, boardwalks, interpret)
- Major new floodplain/aquatic habitat
- Small stone weir to create marsh headpond
- Side ponds dredged to create fish/littoral edge habitat
- Dyking to protect railway and Bayview Avenue from flooding
- Revegetation and habitat improvement of side ravines

River

Channel

- Linear massed and formalized plantings
- New, separated pathway system
- Reconfiguration of fences and R.O.W.s
- Passive recreation and interpretive nodes
- Stairs down from bridges
- Major new green space link between Ataratiri and West Riverdale, west of Bayview extension
- Tilted flushing channel with pool/riffle system
- Further redevelopment to address the river

Portlands

Delta

- Passive recreational/ educational open space
- Bio-physical link; river-lake
- Biological research
- New delta and wetlands habitat
- Peripheral higher density, mixed commercial and industrial 'green industries'

Figure 2.12 *Lower Don strategy plan*

Three distinct landscape units and the peculiar hydrology of the river formed the basis for the long-range restoration strategy. Plates 2.4a, b and c illustrate the dramatic changes in the river landscape as one moves up the valley

Source: *Task Force to Bring Back the Don, 'Bringing Back the Don', Toronto: Hough Stansbury Woodland, Prime Consultants, in association with Gore and Storrie Ltd, Dr Robert Newbury, The Kirkland Partnership, 1991*

Plate 2.5 *The Don floodplain at the 'Rosedale Marshes'*

(Photo: Steve Frost)

river behaviour. As a consequence of glaciation and land history that left an almost flat gradient in the lower river, however, its natural tendency has been to fill with sediments carried down from its upper reaches. These sediments have been continuously dredged from the Keating Channel since the river was channelized in the 1880s to protect adjoining development from flooding. Preliminary studies suggested that to establish the new estuarine marshes at the mouth of the Don required an increase in the gradient of the channel to allow the natural delta-building process in the lower valley to occur at the mouth. However, later hydrological examination showed that a more direct realignment of the river over a shorter distance to the mouth would allow a delta to form within the existing gradient. Thus, over time, the delta could link the river biologically and physically with Lake Ontario, and reintroduce a portion of the original marshes previously destroyed. To realize this objective required an integrated three-part restoration strategy up the river. This included:

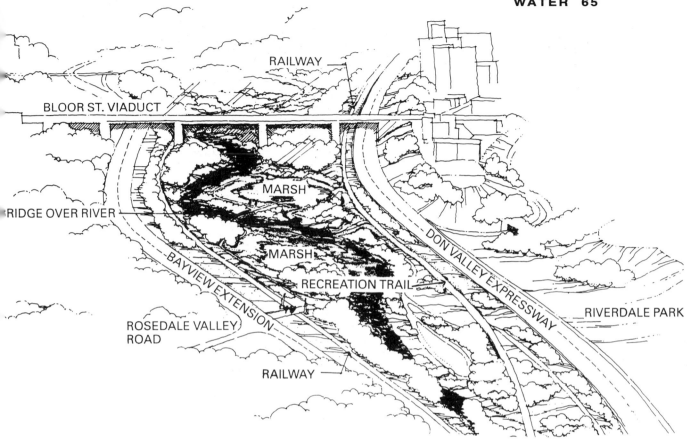

Figure 2.13 *Conceptual sketch from the same location showing the proposed marshed, meadows, walkways and passive recreation*

- the re-creation of a delta and marsh habitats over time from accumulated sediments, where the river meets the lake;
- re-creation of minor meanders and a narrower river within the confines of the existing channel to create fish habitats and naturally spaced pools and riffles (rapids) along its length;
- the creation of upland marshes, ponds and meadows, whose biological function would be to assist in improving water quality, and help balance highs and lows in water flows and flooding.

Thus, the journey up the valley would pass through three distinct kinds of place. The

Plate 2.6 *The channelized Don*

(Photo: Steve Frost)

Figure 2.14 *Conceptual sketch showing tree planting, bike and pedestrian ways and stopping places for **watching the water***

Even with the flat gradient of the river a slight meander is noticeable and can be reinforced with new aquatic planting within the channel, since its width is much wider than that of the natural river channel

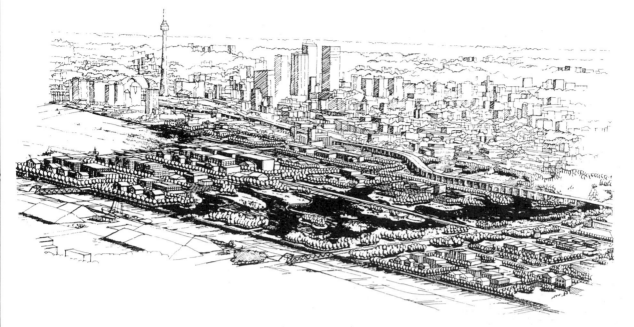

Figure 2.15 *Conceptual sketch showing a potential marsh at the river mouth associated with future urban renewal*

Plate 2.7 *The industrial port and existing mouth of the Don*

river mouth would be integrated with future urban development proposals for the portlands area. Surrounded by mixed 'green' industrial, residential and commercial development the visitor would look over a large open space – a delta and marsh with the river flowing through it to the lake. A research station would monitor the delta's evolution and provide information and exhibits on its research. People would learn about how the river works throughout the watershed, and about the restoration strategies in the Don River. They would also learn how changes to river hydrology can create new and more complex habitats for wildlife, places to spend quiet leisure time and reinstate beauty and dignity to a once degraded valley.

Walking north, the river's character would change from the open and marshy environment of the delta to a somewhat more formal, linear, urban character, with trees lining the water's edge and improving aquatic habitat, pathways for walkers and cyclists, places to stop, picnic, fish and listen to the sounds of running water, and new points of access along the length of the channel.

At the upper reaches of the Don people would enter a third landscape, a place that broadens out into a large floodplain, with marshes, meadows and forested slopes, walks, interpretive stops and picnic places. The floodplain experience would, in turn, change into a fourth landscape – the narrow forested ravine tributaries that make connections with the local communities and parks of the city. Throughout the walks up the Don there would be evidence of Toronto's industrial history: Todmorden Mills, one of the first water-driven sawmills in the valley; the Don Valley Brickworks, where the bricks that built much of old Toronto were once made.

Implementation

The picture presented in this restoration strategy was intended to be bold, imaginative, yet pragmatic in its vision. It also needed to be incremental – staged over many years. The Task Force recognized that implementation was connected, first, with political agendas and funding that involved the formation of partnership agreements between various levels of government and private interests; second, with the adoption of the strategy into the city's official plans and policies; and, third, by small-scale incremental action by local communities. Among many implementation- and management-related strategies were:

- co-operative ventures between the city and other watershed municipalities, senior levels of government, the railways and other private interests and community groups to improve water quality, create habitats, reforest valley slopes and provide access into the valley;
- establishing major open space links in the lower valley through acquisition, easements and related planning instruments available under Ontario's Planning Act;

a

Plate 2.8 *The beginnings of implementation of the Don*

Staircase from the pedestrian bridge linking two parks on either side of the river provides the first access to the bikeway in some 8 kilometres

b

- development of approaches to soil remediation that combine new technologies, policies and issues of legal liability;
- planning strategies that will establish a long-needed identity and value for the valley as a distinct place in the city, having its own planning area and official plan designation that protects valley property in private as well as public ownership;
- the creation of partnership agreements between various levels of government to resolve fractured landownership in the valley necessary to implement restoration strategies.

After the report's completion in August 1990, the political and community impetus for the Don's restoration began to grow. The Metropolitan Toronto and Region Conservation Authority, whose mandate is to protect and manage the watersheds in the Metropolitan Toronto Bio-region, initiated a larger Don Watershed Task Force, a body made up of citizen representatives, experts in various fields, local politicians and government staff. The mandate of this Task Force was to develop a co-ordinated strategy plan for the watershed and oversee its implementation. The Royal Commission on the Future of the Toronto Waterfront also adopted the Lower Don plan as a central component of its overall waterfront and bio-regional planning activities, with support from local municipalities and provincial and federal governments.

At a citizen level, much work was achieved in the first two years. With the co-operation of the City, reforestation of some of the valley's denuded parklands was begun, a staircase was built from one of the bridges crossing the valley linking the two parks on either side of the river and providing the first access to the bikeway in some 8 kilometres of river corridor, and the creation of a demonstration wetland was begun. As an ongoing process of renewal and healing, the Don strategy involved key principles, including a fundamental understanding of process as a biological idea that is also integrated with social, economic and political agendas, economy of means where the most benefits are available for minimum input in energy and effort, and environmental education, where the understanding of nature in cities becomes part of a learning experience that begins with community empowerment and action.

Wastes as valued opportunities

As the restoration of the Don illustrates, water pollution issues are best served when they become part of an integrated design strategy that combines biology and technology, social and economic concerns. It is a fact that sewage treatment technology becomes progressively more complex the more elements that are removed from the water. Primary treatment is a mechanical process of separating used water from the materials that have been added to it. The product is sludge, the total organic and waste-water materials generated by residential and commercial establishments. It is made up of the settled sewage solids combined with varying amounts of water and dissolved material that are removed by screening, sedimentation, chemical precipitation or bacterial digestion..Raw, it is very high in moisture and biologically unstable. When subjected to secondary treatment it is anaerobically digested and produces methane gas and carbon dioxide. The digested material has a high degree of biological stability, and is rich in phosphates, nitrates, potassium and trace elements. Tertiary treatment aims to remove 95 per cent of all remaining substances and chemicals, leaving the water in a drinkable form which is then ready to be returned to the natural system. The technology involved is, however, very costly and few communities are able to afford conventional tertiary treatment systems.

When the products of the treatment plant are seen to be valuable rather than as wastes to be disposed of, then current practice of dumping, by landfill, dumping at sea or incineration, can be regarded as the misuse of resources. In almost all countries where large-scale technologies are limited or expensive to buy, alternative approaches have to be found. It is simply a question of pragmatic long-term investment in a healthy environment. China, as with other Asian countries, has long demonstrated the importance of human wastes to agricultural development and has provided practical approaches for treating them to minimize health hazards.[37] More will be said in Chapter 5 on the relationship of organic urban wastes to the land, particularly its relevance to urban agriculture and the reclamation of urban land. Given that the presence of industrial contaminants in waste water is a problem that must be addressed at source, the concept still provides a basis for solutions to the practical realities of waste disposal in urban areas.

Land as a filter

The advanced treatment and disposal of effluent may be accomplished by returning water and the nutrients it contains to the environment. The soil's micro-biological and chemical capacities are used to filter out nutrients for reuse by plants, and return pure water to ground reserves. Studies at Pennsylvania State University have shown that given the right soil types, agricultural crop yields are usually increased as a result of land application of waste water. Secondary treated domestic waste water has considerable value as fertilizer. As an example, the application of waste water at 5 cm per week during the growing season at the Penn State project provided the equivalent of 232 kg of nitrogen, 245 kg of phosphate and 254 kg of potash per hectare. This would be equal to applying 2.2 tonnes per hectare of a 10–10–11 fertilizer. Over a ten-year period the 5 cm per week

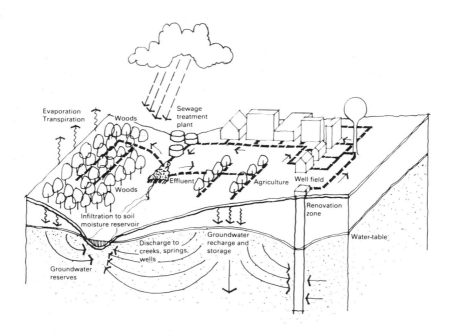

Figure 2.16 *The waste water renovation and conservation cycle*

Land application of waste water, after secondary treatment, is an advanced treatment method. This approach (the 'living filter') considers effluent and nutrients as resources rather than as a product for treatment and disposal. Treatment is provided by natural biological and chemical processes as the water moves through the living filter provided by soil, plants, micro-organisms and related ecosystems. The renovated water then percolates to recharge the groundwater reservoir

Source: *William E. Soper,* Surface Application of Sewage Effluent and Sludge, *Madison, Wisc.: ASA-CSSA-SSSA, 1979*

waste-water application resulted in annual yield increases ranging from 8 to 346 per cent for corn grain, 85 to 191 per cent for red clover and 79 to 139 per cent for alfalfa.[38] Other studies have shown, however, that trace metals percolating through the soil can affect aquifers detrimentally in situations where precipitation is high, and where excessively contaminated effluents are applied to porous soils.[39]

Penn State research has also demonstrated the value of forest land as 'living filters', that have application in rural or other situations where land and appropriate soils are available. Advanced treatment of sewage effluent is provided by natural biological and chemical processes as it moves through the living filter provided by the soil, plants and micro-organisms. Purified water then percolates to recharge water reservoirs. In ten years groundwater was raised by 4.3 metres despite the fact that 9 million litres a day were pumped by the university from wells that also had to supply the rest of the town.[40] In some areas waste water can be applied to the forest all year round, the sheltered humus of the forest floor preventing the freezing that is characteristic of open lands in winter.

Since the early 1960s, waste water has been used in the irrigation of several forest stands consisting of eastern mixed hardwood and red pine. It has been found that satisfactory water purification was obtained in all forests where waste water was applied at a rate of 2.5 cm per week during the growing season, with total annual nitrogen loadings of 150 kg per hectare.[41] The studies also found that the rate of application is critical to the sustainability of the existing forest ecosystem. Overloading may cause the collapse of the system's water renovation capacity. However, they also found that new developing forest stands composed of pioneer species can be very efficient and have a much greater renovation capacity than mature stands. Tree growth and wood fibre production may also be considerably increased. After sixteen years of waste-water irrigation the average diameter of white spruce was 20.3 cm in comparison to 10.1 in the unirrigated control forest. The average height of trees in the control forest was 4.5 metres, whereas irrigated trees had an average height of 9.2 metres.[42]

As current liabilities of waste-water disposal begin to shift focus to the benefits of nutrient recycling, the application of waste water to forests and agricultural land provides a viable alternative. At the same time, considerable research since the 1960s has shown that these nutrients must be applied at rates and quantities that can be assimilated by the land and vegetation. Appropriate bio-regional conditions, pre-treatment and comprehensive monitoring programmes are necessary if the overloading of the system is to be avoided.[43]

Aquatic plants as a filter

Biologists have known for many years that marshes have a very high capacity for recycling wastes. The highly productive nature of marshland ecology promotes the uptake of nitrates and phosphates by aquatic plants. Research in Mississippi has shown that, annually, 0.4 ha of water hyacinths absorbs 1,600 kg of nitrogen, 360 kg of phosphorus, 12,300 kg of phenols and 43 kg of highly toxic trace metals.[44] A variety of wetland and solar aquatic systems were developed experimentally for commercial purposes in the 1980s and 1990s. Columbia, Missouri, has contracted for a 36 ha sewage treating wetland and over 150 wetland treatment systems serve cities and towns in the USA.[45] Harwich, Massachusetts, and Providence, Rhode Island, have installed solar aquatics systems, and the Providence facility is intended to process between 75,000 and 113,000 litres per day of domestic and industrial waste.[46] Other smaller systems have been installed in such places as Muncie, Indiana and the Toronto Board of Education

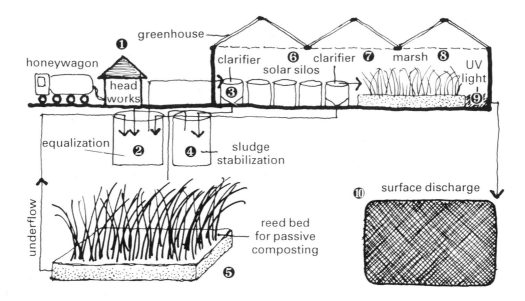

Figure 2.17 *Solar aquatics septage treatment system*

As a typical SAS septage treatment plant, effluent is discharged into a headworks (1) consisting of a receiving station and degritter. From there it flows by gravity into in-ground tanks (2) where the effluent is blended to compensate for variations in loads and is preconditioned. Following equalization and preconditioning, effluent is pumped to a clarifier (3). From here sludge is pumped to the sludge stabilization tank (4) where it is aerobically stabilized. It then flows to the reed bed (5) for passive composting. Underflow from the reed bed is recycled by pumping it back into the equalization tank (2) for further treatment.

Liquid from the clarifier enters a series of solar silos (6) that contain the plants and animals involved in the clean-up operation. After moving through the silos by gravity flow, the liquid enters a second clarifier (7). Solids from this clarifier are recycled back to the equalization tank (2), where they re-enter the system, reseeding microbes as they travel.

Next, the liquid moves through a gravel-filled de-nitrifying marsh (8) planted with grasses and is then sterilized with ultraviolet light (9) before being discharged (10). Depending on total suspended solids limits, a sand filter may also be required at this point in the system for final polishing before purified water is discharged

Source: *Ecological Engineering Associates, 13 Marcom Lane, Marion, MA, USA*

Nature Interpretive Centre at the Boyne Conservation Area, Ontario. In cold climates these systems must be contained in greenhouses. In warm sunny regions such as the Middle East, or the south-west US they can operate in the open. Conventional waste-water plants release large quantities of excess nutrients and toxic chemicals into city sewers, lakes and oceans. They also create millions of tonnes of sludge each year, much of which is loaded with heavy metals. Aquatic systems, however, suspend solids by aeration, and use solar energy, bacteria, plants and animals in a system designed to enhance natural responses which bio-chemically change or remove contaminants from sewage. Taking ten days to process, waste water circulates through a series of large transparent plastic tanks or silos, each of which are a functioning ecosystem. Bacteria and algae use sunlight to digest organic matter suspended in the liquid, converting ammonia to nitrite, then to nitrate. Water plants and algae consume nitrates and zooplanckton and snails feed on algae. Subsequent steps include engineered marshes with plants that take up heavy metals and other toxic chemicals; other silos contain fish that feed on nitrate-eating phytoplankton, and water hyacinth take up other heavy metals. The final step are 'polishing' marshes from which water flows clear and sludge-free.

Solar aquatic systems require much less land than marshland treatment and are no more land-consuming than conventional treatment plants. The operating cost is also about two-thirds the cost of conventional treatment.[47] In addition, waste water is seen as a resource rather than a waste to be disposed of in the larger environment, and there is potential revenue to be gained from hydroponically-grown flowers and plants. Some problems remain, however. Many heavy metals, such as lead and cadmium, stored in plants must be periodically gathered and disposed of, a limitation that is, none the less, still a cheaper and safer option than current methods of incineration or burying tonnes of metal-laden sewage sludge in landfill sites from where it is released into the larger environment. Government agency resistance to adopting such systems is also inevitable since their operation requires a very different type of expertise. There is a need to adapt these natural systems to local conditions and to ensure the control of input, facts that makes them 'somewhat suspect in the highly standardized world of water quality engineering'.[48] In summary, however, as Bob Bastian of the US Environmental Protection Agency has indicated, 'much of the future of alternative treatment will depend on whether regulatory agencies push for compliance with strict water quality standards'.[49]

Water purification using vegetated wetlands is also gaining recognition in the Netherlands, Germany and other European countries. There are some thirty wetland projects in the Netherlands dating from the 1960s. Some include camping sites where waste water is purified on a small scale, others have been built by water boards, municipalities and industry to treat the effluent from treatment plants.[50] In the Dutch government's fourth National Report of Physical Planning (1988, updated 1990) it was suggested that vegetated wetlands should be incorporated in city and land use planning. It proposes, where space is available, the decentralization of smaller purification systems that can serve multiple functions. Studies have proven, in fact, that 'the combination of water purification, nature development, recreation, and agriculture is possible'.[51]

Plate 2.9 *Retention ponds in Kitchener, Ontario*

a *Permanent pond that acts as flood prevention and provides a habitat for fish, wildlife and a natural setting for the surrounding community*

(Source: Kitchener Parks and Recreation Department)

b *Pond in a new housing estate*

Originally engineered as a sterile turfed basin to fill then empty, this part of the storm drainage system was restored and naturalized

(Source: Kitchener Parks and Recreation Department, Hough Stansbury Woodland)

Rainwater – the principle of storage

The basic lesson that nature provides in the water cycle is one of storage. Natural floodplains and lakes are the storage reservoirs of rivers that reduce the magnitude of peaks downstream, by spreading and equalizing flows over a longer period of time. Vegetated soils and woodlands provide storage by trapping and percolating water through the ground with minimum run-off and maximum benefit to groundwater recharge. Water quality is enhanced by vegetation and storage which in turn will contribute to the diversity of natural and human habitat. Thus storm drainage must be designed to correspond as closely as possible to natural patterns, allowing water to be retained and absorbed into the soil at a similar rate to natural conditions. This principle is now well recognized in western countries as the realistic alternative to current practice. A discussion of a few of these alternatives is appropriate here.

Groundwater recharge

Where the porosity of the soil permits, rainwater applied direct to the land helps to replenish groundwater reserves. Natural drainage over turfed or vegetated land is very useful in controlling and managing stormwater. Functionally it aids natural infiltration into the ground and controls the velocity of water flow, which is essential to control erosion and sedimentation. The objective is to achieve a rate of water run-off that is equivalent to the pre-development levels, helping to minimize flood and erosion damage. Many low- to medium-density housing developments in North America have adopted the practice of overland flow from developed surfaces over turfed areas and channels as an alternative to traditional storm sewers, since it has been shown to be more economical and more beneficial to water quality. Vegetation is the crucial factor that ensures that water is recycled back to the natural system. Natural drainage, woodlands, rough grass and shrubs and small wetland areas provide the functional basis for determining open space patterns as the new town of Woodlands, Texas, has demonstrated. This example will be examined on pp. 83–9.

Retention ponds and lakes

Retention ponds are, when modelled on nature, a means of controlling water run-off by modifying flows, smoothing out peak loads by releasing water slowly to streams to lessen the danger of downstream flooding. Temporary ponds also help to replenish natural groundwater where soil porosity is high and where severely contaminated urban run-off is not an issue. Permanent ponds are effective in enhancing water quality. Pollutants entering the cycle from rain, roads, paved surfaces and rooftops include a range of organic and chemical compounds and heat from paved surfaces, making stormwater hostile to aquatic life. Since most pollutants in stormwater are attached to sediment particles, allowing these sediments to settle to the bottom of the pond over a period of days, rather than letting them spread throughout the system, is an important method for treating and improving water quality. Vegetation is also a significant factor in stormwater management since trees, shrubs and aquatic species contribute to cooling through shading and water quality before the water enters streams and lakes.

Storage of water is created naturally as lakes, ponds and marshes, and they often occur by accident in many urban places that do not come under the label of parks and playgrounds. The wealth of abandoned industrial or mining lands, vacant lots, waterfront sites and highway interchanges perform, quite fortuitously, a valuable hydrological function by retaining and storing water. This is often lost when development occurs and sites are 'improved'. Storage of water on low- and medium-density development is becoming an accepted water management alternative in situations where land is available for temporary ponds and as an inexpensive and environmentally appropriate approach to urban drainage. The Ministry of the Environment for Ontario has concluded, for instance, that water-retention facilities to control run-off must be incorporated into urban development.[52] Many alternatives are possible, each depending on the nature of the place, its rainfall, topography, drainage patterns, plant cover and soils and type of development.

Permanent storage

This is appropriate where a continuous supply is available and where inflow and outflow and soils permit stable conditions. Ponds that can be maintained at some minimum water level offer multi-purpose potential for community uses. Balanced ecosystems of plants, animals, fish and other aquatic organism maintain stability, provide places for nature study, education and fish management. Open water provides places for boating and recreation and visual enhancement. Examples of the concept have been implemented in a variety of housing developments in the US and elsewhere.

Temporary storage

It might well be assumed that densely urbanized areas do not have the capability to cope with rainfall storage. It was for this reason, in fact, that storm drainage was introduced in the nineteenth century. Where space is at a premium, or permanent ponds are inappropriate, or where an existing storm drainage system is subjected to additional loads, the principle of delayed return to the receiving body may be put into practice. The floodplains of rivers and streams work on this principle, releasing excess water slowly and smoothing out peak loads. In the city temporary storage is useful in situations where various functions must be accommodated on the same area of ground. It can be designed to accumulate water during rainstorms and drain completely after the storm over a period of time. The land thus serves dual purposes; assisting hydrological functions, but still providing space for various other uses. Golf courses, playing-fields, cemeteries and parks are typical situations where management could accommodate compatible uses. In fact, they often do by default rather than by design, as the flooded fields and lawns of many urban parks after sudden rain will testify.

Where non-paved areas are insufficient or unable to cope with natural storage, many other types of open space are potentially available that can fulfil hydrological roles. A 1971 study, for instance, illustrated how much unbuilt land can exist in cities, with over 57 per cent occurring in downtown areas,[53] and considerably more in suburban areas. Thus, a major proportion of the city's paved or impervious space is, in fact, available and may be designed to accommodate temporary storage. Water may be held long enough to reduce peak flows, then released slowly into storm systems. For instance, parking lots take up a large proportion of most downtown areas. If designed with a hydrological function in mind, such spaces can provide appropriate storage which would not significantly interfere with their parking uses. As an example, it is accepted practice to design parking areas in cold climates for snow storage. A similar practice could be extended into summer months for rainwater. Streets, which may take up to 27 per cent of downtown space, may also act as temporary storage, as may storm sewers themselves. There are many other realistic ways of reconnecting urban water with the hydrological cycle. They include a return to the rainwater butt that stores water off roofs and provides for garden irrigation and washing cars, thus conserving drinking-water supplies; discharging water on to vegetated ditches, lawns and backyards to collect in shallow depressions, a strategy that has been implemented with considerable success in the Village Homes community in Davis, California.

Roof run-off storage

Flat roofs and basement storage are other places that can be used to serve a storage purpose on the principle that buildings should take care of their own rainwater rather than throwing the problem on to the public domain. Flat roofs performed this function until the introduction of the 'upside down' roof which in modern building is designed to remain free of water at all times. Many older roofs do still pond considerable amounts of water, by default rather than by design, and policies in many municipalities encourage the practice where there is insufficient land. As previously discussed, collecting water from domestic rooftops with the old fashioned rainwater butt was common practice in most places until mains water was introduced. Again, the universal acceptance of the storm sewer has in most cities replaced this conservation-minded practice. In country areas rainwater is relatively pure and can be used for most purposes. In the city, rain falling through the haze of urban pollution picks up impurities that makes it unsuitable for some things, unless it is treated first. It is quite acceptable, however, for uses such as washing cars and windows and watering the garden. Collected in a butt, it also helps to conserve domestic water supplies that are currently used for every function.

Wetlands for stormwater purification

In addition to their function as providers of biological treatment for waste water, discussed on pp. 73–5, constructed wetlands have become an increasingly important method of enhancing the quality of run-off water in cities in Europe, the UK, the US and Canada. They also have immense importance in controlling stream flows and downstream erosion, and restoring impaired habitats for aquatic and terrestrial species. While there is considerable variability in the efficiency of wetlands at removing contaminants, it has been shown that they can remove 70 per cent of excess nutrients, and destroy bacteria and viruses. Heavy metals may be accumulated either in sediments, or are associated with organic materials, and many pesticides, oil and grease are broken down by microbes and plants. They have also been developed to treat the acidic water quality of coalmining and ore-processing plants. For instance, it was found that wetlands created in western Pennsylvania to treat water from old mine sites resulted in improved water quality as well as furnishing habitat for birds, mammals, reptiles and amphibians.[54]

As discussed on pp. 77–8, the chemical and physical mechanisms for treating stormwater include dispersal by evaporation, sedimentation and absorption where dissolved pollutants adhere to suspended solids and settle out, and filtration through vegetation and soils. Since studies have shown that the first 2.5 centimetres of rainfall carries 90 per cent of the pollution load in a rainstorm,[55] the design often includes a pre-treatment detention pond to collect and settle out sediments on which much of the pollutants are carried, before the water flows into the wetland. Like all natural systems, each place is unique and requires its own particular adaptations.

SOME CONSIDERATIONS OF DESIGN

The basis for design form

In this review of water I have tried to make connections between urban and natural processes. It becomes apparent that if we examine these connections in the context of the urban environment, the city's open spaces become a fundamental factor in re-establishing hydrological balance. City spaces have a significance beyond the transportation, economic or recreational assets that we normally ascribe to them. The notion of investment in the land now begins to acquire conservation and health values. An ecological basis for urban form suggests that when the city's water is recycled back into the system there are reduced costs and increased benefits. Urban development becomes a participant in the workings of natural systems. The lands associated with sewage treatment plants that are currently underused or unproductive can, where land areas permit, be made to function as groundwater recharge areas and help in purification of the city's waste water. Urban forests and market gardens acting to purify waste water and recharge groundwater can also become valued when they provide food, timber and wildlife reserves close to home. The value of urban forests to the city is immense and provides the basis for useful, productive and low-maintenance landscapes. The city's residential parks, open spaces and waste lands, parking lots, playgrounds and roofs could be adapted, where appropriate, to serve a hydrological function by creating temporary or permanent storage areas and wetlands, thereby helping to redress the problems of erosion and pollution for which many cities have initiated policies. Similarly, allotment and market gardens, within and on the edges of the city, provide the basis for commercial and recreational land uses that serve productive and environmental functions and increase landscape diversity. These will discussed more fully in later chapters.

Design and the new symbolism

Thus far I have dealt with the ecological and functional basis for appropriate design and management that may achieve a measure of environmental sustainability. It will be obvious, however, that water, perhaps more than any other element of the landscape, has deep-rooted spiritual and symbolic meanings to which design must respond. As an element of great experiential power, water has historically been manipulated and shaped to create places of delight and beauty. It has reflected cultural attitudes towards nature. The Romans celebrated water in their engineering and architecture. The medieval gargoyle capitalized on the inherent opportunities to celebrate water falling off cathedral roofs. The exuberant splendour of the Italian water garden, created by the volume, light and sound of its fountains, exploited the hillsides and streams around Florence and Rome by using natural gravity. The Japanese miniature garden symbolized in miniature the oneness of nature and culture. The placid lakes of the English landscape garden expressed the romantic and pastoral qualities of England's rural landscapes.

The task today is to create a new design symbolism for water (and urban natural systems as a whole) that reflects the hydrological processes of the city; an urban design language that re-establishes its identity with life processes. The opportunity for this to

Figure 2.18 *The symbolism of the great gardens of history*

As an element of design, water was manipulated to create places of delight and beauty. Operating without electricity, their form tended to be an expression and symbol of the place, gravity and natural opportunities available

a Italy

b England

occur lies in the establishment of a vernacular landscape whose aesthetic rests on three factors. First, on its ecological and functional basis for form. Second, on the *integration* of design objectives where design becomes multi-faceted and experiential. Single-purpose solutions to problems tend to create other problems. Third, and most important, on the notion of visibility (see some design principles, p. 30). The motivating forces that have shaped the city's conventional systems and technologies have been based on the concealment of the very processes that support it. Revealing and enriching the processes of nature and the diversity of the city's cultural landscape, therefore, lies at the heart of the urban experience and artistic form.

Expressions of this notion of visibility from European and North American cities illustrate the opportunities. The engineering functionalism of the sewage treatment plant can be brought into harmony with form and symbolized in unexpected ways, as in the flowform sculptures at Jarne, Sweden.[56] Here sewage water cascades down various sculptured basins and is aerated as it drops. Porous paving materials that permit water to penetrate through to the soil and ground vegetation to grow while maintaining a hard surface, offer other opportunities for integrating design with hydrology. In Germany, the variety and pattern of ground surfacing, the channels that direct water movement and ponding and the beauty these add to the city floorscape are matched by the hydrological and climatic functions they perform; a welcome relief from the visual tyranny and sensory deprivation of asphalt surfaces. In Canada, art can capture, as part of the city's storm sewer system, a moment in the water cycle.[57] Fish have been painted on city catchbasins to remind people of where the water goes, and on Toronto's Don River, cutouts of birds and animals that once inhabited the valley have been placed on the chainlink fences that now separate walkers and cyclists from the valley's railway corridor.

Woodlands and hydrology

While we are concerned here with existing cities and their rehabilitation, the new city of Woodlands, a mixed-use community 40 kilometres north of Houston, Texas,[58] provides an excellent example of urban form adapting to hydrological processes.[59] It is relevant particularly to low-density suburban areas where it is possible to adapt the natural conditions of a site to urban development. Located in a subtropical region that has mild winters and warm, humid summers, the key features of the area are its extensive mixed woodland, flat topography and imperfectly drained soils. Heavy seasonal rains result in frequent flooding of streams. The problem facing the planners was how to preserve the woodland environment while permitting land drainage for urban development. Much of the site is poorly drained, a condition that results in standing water after rains. Streams have very low base flows and shallow floodplains in the flat topography. Thus the

Figure 2.19 *Making visible the processes that sustain life: sculpture and sewage works*

The flowform cascades at the Rudolf Steiner Seminariat, Jarne, Sweden. The cascades aerate effluent from a community of more than 200 people

Source: *Peter Bunyard, 'Sewage Treatment in a Swedish Sculpture Garden',* Ecologist, *no. 1, Jan.–Feb. 1978*

environmental analysis carried out in the initial stages of the general plan clearly showed that maintaining hydrological balance was a key planning factor in which the preservation of permeable soils played a major role. It also became apparent that the introduction of conventional piped storm drainage would destroy much of the forest landscape. Run-off had to be retarded to ensure maximum recharge.

In response to these factors, the development was shaped to respond to hydrological determinants. Planning objectives included design for maximum recharge, protection of permeable soils, retardation of erosion and siltation, increase of stream baseflow and protection of vegetation and wildlife habitats. Roads were located on ridges away from drainage areas, intensive development was located on impermeable soils, minor residential streets were oriented perpendicular to slopes and used as berms to impede flow over very permeable soils, porous paving was recommended for paved areas to increase water storage, the existing natural drainage of streams, draws and ponds was used, and artificial impoundments and settlement ponds were employed to enhance this network. The open space network was related to the drainage system and vegetation.

At the town scale, primary open spaces were conceived as a conservation zone, incorporating areas performing valuable hydrological vegetation and wildlife functions.

Plate 2.10 *Children and play: sculptured stormwater flows in a German housing development*

(Photo: Herbert Dreiseitl)

Plate 2.11 *Artistic expression of the catchbasin and drainage swales*

(Source: City of Toronto Parks Department)

Plate 2.12 *Retention pond and golf course, Woodlands, Texas*

(Photo: Tom Coyle)

Figure 2.20 *Site planning principles: hydrology*

Removal of forest cover and the addition of impervious surfaces will increase frequency of flooding and change stream characteristics. However, the use of a natural drainage system increases water–table recharge and increases lag time for run-off entering streams. Protection of floodplains and drainage swales is imperative for flood control, for regulating stream control and for maintaining water quality

Source: *After Ian L. McHarg and Johnathan Sutton, 'Ecological Plumbing for the Texas Coastal Plain.' In Grady Clay (ed.)* Water and the Landscape, *New York: McGraw Hill, 1979*

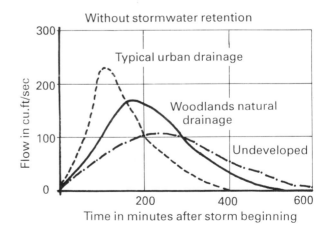

They included the 25-, 50- and portions of the 100-year floodplains of existing streams. Secondary open space included other components of the natural drainage system, secondary draws, impoundment sites and recharge areas. Tertiary open space at the scale of individual development areas included vegetation not cleared for building, small green spaces among detached houses, vegetation buffers along roads and so on.

Analysis of soil types resulted in the designation of a 'permissible coverage' factor. This defined the percentage of an area that may be made impervious without affecting the absorption capability of the soil after a high frequency storm (i.e., 2.5 centimetres in 6 hours). For instance, highly permeable 'A' type soils that cannot be used to drain upslope less permeable soils have a high permissible coverage factor. The permissible coverage factor provides a simple indication of the maximum density of allowable development, based on soil performance characteristics.

A similar performance measure, 'permissible clearance', was formulated for different categories of site vegetation. It provided an indication of the percentage of an area that might be cleared without serious environmental disruption. Determination of the clearance factor was based on the role of various types of vegetation in maintaining hydrological balance, tolerance of species to disturbance and their value as wildlife habitat. From the combination of these two sets of factors, a 'landscape tolerance' was derived as an indicator of development potential.

Following the development of a general plan, site planning guidelines were prepared which provided a means of communicating ecological principles to planners and designers. Their purpose was to permit identification of areas intrinsically suited to specific land uses – residential circulation and open space – in the context of a natural drainage system and the maintenance of a woodland environment.

As construction phases proceeded it was evident that the natural drainage system was effective. It was reported that despite 33 centimetres of rain in three days and 10 centimetres in 1 hour, no surface water remained after 6 hours. Ponds filled and returned to normal within the same period of time. While estimates have shown that the effective costs of the natural drainage system are, in fact, equivalent to, or only slightly less than, conventional piped systems, they none the less represent other kinds of economies. It was pointed out that conventional developers who show little concern for environmental values can realize a greater return on investment than the Woodlands new town is obtaining.[60] Part of this profit is gained, however, at public expense – the costs of solving flooding, erosion and water quality problems being transferred to public authorities. With the Woodlands project, however, the cost of initial prevention is substantially less than the costs of corrective measures after development.

Objectives	Adaptations	
Reduce flooding.	Ensure ability of existing primary and secondary drainage channels to handle storm run-off by defining drainage easements. These drainage easements will be determined by the 25 year floodplain, however, a minimum vegetation easement of undisturbed forest and understorey must be respected: 90 m for primary drainage channels and 30 m for secondary drainage channels.	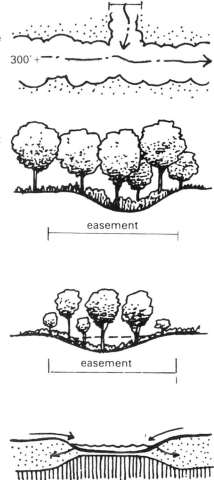
Minimise erosion and siltation.	Prohibit clearing of ground cover, shrub understorey, or trees within drainage easements.	
	Enhance existing channels where necessary with berms and 'create' natural swales by introducing layered plantings of native vegetation.	
Contribute no increase in off-site discharge during the Design Storm.	Provide adequate storage of run-off generated by Design Storm in impoundments or temporary water storage ponds.	
Retard run-off and maximize recharge to even base flow of streams.	Use check dams in swales and on lots to slow flow over permeable soils to enhance recharge.	
	Install trickle tubes in impounded areas to permit even flow.	

Figure 2.21   *This figure illustrates a few of the principles developed to preserve natural drainage systems and siting of housing and roads*

Source: *Wallace, Roberts and Todd, 'Woodlands New Community, Guidelines for Site Planning Report', 1973, Philadelphia*

AB soils (LA,EU,BL,BOH,AL,WI) which receive no run-off from other soils

These soils have a high storage capacity for excess run-off, but when located upslope from less permeable soils, they cannot be used to drain them. In this condition, the excess storage capacity should be used to recharge run-off from higher density development on the AB soils themselves.

A soils have more excess storage capacity than B soils. Therefore, less area of A soils is required to recharge a given amount of run-off.

Excess storage capacity

A soils   B soils

Management Guidelines

A soils may be cleared up to 90% and still achieve local recharge of the 2.5 cm storm.

B soils may be cleared up to 75% and still achieve local recharge of the 2.5 cm storm.

Areas used for recharge should remain wooded.

A soils
Up to 90% cleared and rendered impermeable

B soils
Up to 75% cleared and rendered impermeable

Housing Suitability

On A soils all types and densities are suitable.

On B soils most types and densities are suitable. Housing types and densities which require more than 75% clearance cannot be accommodated without additional uncleared A or B soils. This would result in decreased gross density.

Siting Considerations

Situate buildings and impervious surfaces on higher elevations so that run-off will drain to lower elevations where it can be recharged.

Situate buildings and impervious surfaces so that they drain to the uncleared area.

Hydrology and the inner city

An attempt to give physical form to the inherent potential of rainwater was made in the early 1980s in a housing project in central Ottawa. Called the LeBreton Flats Demonstration Project, it was a joint undertaking by the National Capital Commission, Canada Mortgage and Housing Corporation and the City of Ottawa. The overall project aimed to demonstrate innovation in housing, responding, first, to the social and economic realities of the city and, second, to the need for conserving energy in housing design. Located on the edges of an established neighbourhood in the inner city, the project was concerned with the rehabilitation of derelict urban land within an existing medium-density community. Part of the new housing was grouped around a small 0.5 hectare park, the focus of the entire neighbourhood. It became apparent, after evaluating local community concerns and interests, the resources of the site and the project's environmental goals, that these must extend to the park itself. It was apparent that the concept of stormwater retention should be explored in medium-density residential areas of the inner city. Since the rainfall of the Ottawa region typically falls as sudden storms, the environmental problems of conventional urban drainage are equal to, if not more severe than, those of outlying suburban areas. The potential opportunities were twofold:[61]

- to demonstrate on a small scale the creative and practical alternatives to traditional site development practice in storm drainage in an inner-city park; and
- to create a place that would integrate environmental with educational objectives, i.e., storage and slow release of urban drainage to create temporary ponds for play and recreation activities that are dynamically tied to the hydrological cycle.

Several important considerations dictated the design solution. The first was the smallness of the park in relation to the size of the community using it. The second was the volume of water that could be accommodated during a storm while not interfering with other essential park functions and ensuring safety. The third was the community's perception of the kind of park they needed. During many intensive meetings with both adults and children it was apparent that there was a strong desire for a 'green' park with grass and flowers, places for sitting and strolling, for meeting and informal games. This was understandable in view of the dearth of parks in the general neighbourhood. In addition, it was evident that most active games requiring hard surfaces were played on local streets, which, like most established neighbourhoods, traditionally have been the social hub for both children and adults.

The way the hydrological functions of the park were expressed as design form was directly affected by these considerations. The need for a green park permitted most of the

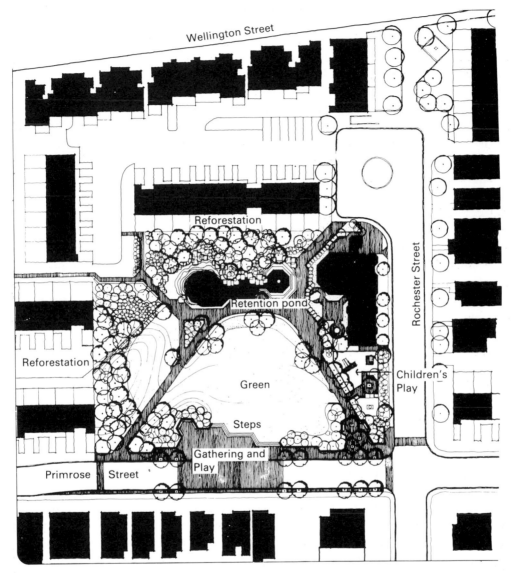

LeBreton Park and surrounding residential housing

Figure 2.22 *Plan of LeBreton Park, Ottawa, and surrounding residential housing*

Source: *Hough Stansbury Michalski, Toronto*

park to be soft – grass, trees and planting beds. At the same time the need for socializing, gathering and informal hard-surface games resulted in a new approach to the street design bordering the south side of the park. Children's activities were recognized as a valid function of local streets. By incorporating safety measures to slow local traffic, the park concept could integrate both the street and the park as one unit, the street providing hard surfaces, the park providing the soft ones. The ability of the park to serve a hydrological function was thus enhanced by absorbing run-off to the fullest extent over its vegetated surfaces and by the inclusion of a stormwater detention pond. This was hard-surfaced, since the notion of a naturalized impoundment with a mud bottom and aquatic plants was rejected as impractical.

The pond was designed to hold 76,000 litres, which is the maximum amount of water that might occur in any one rainstorm in a two-year period. A peak rainfall of 15 minutes, amounting to 2.3 centimetres for the Ottawa region, was assumed for purposes of calculation. Roof water obtained from adjacent housing and overland flow within the park was collected by a series of drainage channels that deliver the water to the storage basin. Water from parking areas and streets in the housing project was not collected to minimize the introduction of pollutants in a general play environment. A small adjustable weir and catchbasin inlet were designed to permit the pond to fill to a depth of 46 centimetres and drain out slowly to the storm sewer over a period of time. An average of two to three days was regarded as a reasonable period for a temporary storage situation.

The character of the pond would, therefore, be constantly changing: becoming a potential ice surface in winter, and filling periodically after rain during the summer. Its hard surface was designed to take intensive use and was contoured to permit many different activities, depending on whether it was wet or dry. Thus children's play could respond to the changing environment: focusing on water play after a rainstorm and dry weather activities when the pond dried out.

The purpose of this experiment was to integrate an urban natural process – water and its management – with social and education values. It involved natural process and design in the familiar and daily routine of city living. A hydrological function was established for park settings that had traditionally been devoted to recreation alone. And hydrological functions became part of the understanding and experience of the park and of the neighbourhood. In these ways it was intended to bring together solutions to environmental problems in the city with social needs and art. It was intended to delight the senses, provide a place for creative play, function practically and bring people closer to the continuum of natural events. As I remarked in the first chapter, children learn about life and their environment less by the occasional visit to the nature centre or the museum than by constant and direct experience in their daily surroundings and activities. A pond in the park after a rainstorm that one can splash through is the focus for play and learning

Plate 2.13 *LeBreton Flats, Ottawa: the retention pond flooded*

During a rainstorm the pond fills to 46 centimetres and takes on a very different character inviting unstructured water-related play and visual variety

Plate 2.14 *The retention pond dry*

Flexipave surface provides a durable topcoat for a variety of dry weather activities

while it lasts. It provides the best opportunity for understanding hydrology in cities. Through play children are brought closer to the cycles of rain and sunshine. Water as the ageless material of design can assume a new relevance and it is on this basis that an alternative urban form must emerge.

Some notes in hindsight

During the ten years following completion of the project in 1980, it became apparent that the realities of its evolution were very different from those imagined during its conceptual development. It is here that hindsight provides lessons for what works and what does not in terms of the role of government, community change and park policies.

Things that succeeded

Shortly after completion of the park the detention pond began to be used in interesting ways. Children flocked to the park after rain and used the water-filled pond as a paddling pool, for sailing small boats and for those creative water activities known only to children. During dry periods its surface became a place for other happenings, including biking, roller-skating, tag. One small girl said in conversation that she did not live in the neighbourhood but came there to skateboard every day because it was the best shaped surface she knew of. Adults sunned themselves on the south-facing grassy slopes, flowers were planted, mothers brought their small children to the playground and people talked about the different things they could do in the park.

Things that failed

A number of interrelated events contributed to the failure of the park's original intent. First, upon its completion responsibility for the project was relinquished by the initiating federal authority, and ongoing management became the responsibility of the City Parks Department that had had some involvement in the project's development, but whose park management policies, based on limited budgets, were geared to conventional turf maintenance and occasional litter clean-up. Maintenance staff had no experience in the management of the detention pond which quickly began to fill with debris and broken glass, and which became a hazard for children. Second, people in the community who had worked with the design team to create the park and who had decided to live in the new housing began to move out for a variety of personal reasons, while others moved in. Over a period of years, therefore, fewer and fewer people who knew the reasons behind the park's design and who were able to communicate their knowledge to newcomers, remained. Acts of vandalism, often by older children, began to occur, such as destruction of trees, lighting and benches. Few of the plants were replaced and a sense that the park lacked ownership began to grow. Third, the channels supplying water to the pond were located inside backyard boundary fences rather than on public property. As people altered their gardens to suit their own needs drainage courses were dammed up and the pond began to lose its retention functions during rains and has ceased to flood. A visit I paid to the project in June 1994 confirmed this situation. While some of its main features remained intact, people had no idea that the pond area had hydrological or educational purposes.

•

Interesting lessons can be learned from this experience both from the point of view of policy and design. First, building a sense of community in a new neighbourhood is an ongoing process that must be nurtured through appropriate social mechanisms. These might include a continuing involvement in the project between the development agency and the community that lives there through local community leaders. Neither the federal authority or the city had the mandate or the available budgets to finance such initiatives. At the same time, a rough calculation revealed that the cost of replacing damaged property caused by vandalism in the park would have made a substantial contribution to the cost of hiring a resident community leader, and thus helping to avoid vandalism in the first place. The situation ignores the social and environmental links that need to be made in the evolutionary processes that dictate the development of human and non-human communities. It also raises related issues, discussed in Chapter 3, of how the sense of ownership and control of space can be developed in the design of human communities and the role of the designer in that process.

Thus far, we have explored the role of water in nature under natural and urban conditions. We have seen how developments in supply and disposal technology, by improving human health, have been a central factor in urban growth. But the benefits of health and urban growth have been achieved at the expense of disturbed natural cycles and the creation of a general environmental deterioration with respect to a worsening climate, water pollution and diminished wildlife habitats. By themselves, the technological remedies designed to mitigate some of these problems have only shifted them from one place to another – from the city where they began to the larger environment.

It becomes clear that the opportunities for alternative solutions lie in a better understanding of the nature of the places we live in. Urban ecology provides the conceptual vehicle for city design, for which the principles invoke a basic shift in values. We begin to see wastes as opportunities that can contribute to environmental health and diversity, drawing maximum benefits from the means available. Waste water returned to the land can enrich the soil, provide nutrients for produce, crops and urban forests, giving biological, social and civic value to currently wasted urban land. Stormwater retained in the city's open spaces contributes to the restoration of the hydrological balance and can help ameliorate urban climate. It enriches the potential for integrating environmental, social and aesthetic benefits within parks and urban spaces and brings nature's processes closer to everyday life. And while many specific issues remain to be solved in this process, the fundamental basis for this alternative view remains true.

It has also become clear in our discussions so far that water, being a part of the whole interconnected system of natural process, affects every aspect of the subject at hand. It is central to the maintenance of biological communities – plants and animals – which are themselves vital to the city's environmental health. This chapter has shown how plants are directly connected to the hydrological cycle. They have from time immemorial shared pride of place in civic design. It is now time to turn to plants themselves, since they reveal other issues that must be examined in our search for an expression of the city as a place in balance with nature.

3

PLANTS

•

INTRODUCTION

The inquiring observer of plants in the city may be struck by the extent to which they depend on horticultural and technical props for their survival and health. That individual may wonder what motivates people to use plants the way they do and what purposes they serve; pollarded lindens at the base of thirty-storey tower blocks, tropical plants crammed into the dark and unused recesses of an office interior, the unbroken turf and isolated trees of every city park. Why do people put so much energy and effort into the nurture of cultivated and fragile landscapes that are usually far less diverse, vigorous and interesting than the 'weedy' landscapes that flourish in every unattended corner of the city? Why indeed is the one tended with such care and attention and the other ignored or vigorously suppressed?

My task here is not to provide another guide to design with plants in the conventional sense of discussing principles of form, space, colour, texture and so on. This has been, and continues to be, done very successfully. My intent is to consider plants from other perspectives, to seek a valid basis for aesthetics that has its roots in urban ecology, to explore functions and opportunities for urban plants that are consistent with the ideals of an ecologically sustainable philosophy, and to examine city spaces in relation to their functions and patterns of human behaviour. To this end we must first review briefly some aspects of natural processes, how these are altered in the city environment, how attitudes and perceptions have shaped the city landscape and what opportunities exist for alternative ways of using plants in cities.

NATURAL PROCESSES

Plants are the basis for life on earth. They produce all the oxygen in the earth's atmosphere; they provide the food and habitat through photosynthesis that supports all living creatures. Plants are environment specific, evolving different forms and communities that have adapted to specific climates, rainfall, soils and physiographic types. These major plant communities are grouped into general regions which range from Arctic tundra, northern coniferous forests, temperate deciduous forests, tropical regions, savanna, grassland and desert. Our concern here is primarily with the temperate-zone deciduous forests of eastern North America, Britain and Europe. It is in these regions that the majority of industrialized cities have evolved.

Succession

Each forest type goes through a period of infancy, youth, maturity, old age and rebirth. In some forests such as the northern boreal regions, this process of succession is dependent on fire that reduces them to ashes, after which rebirth starts again immediately. In the deciduous regions the process is continuous under natural conditions. Parts of the community die and regenerate, but the forest as a whole remains. Odum describes this constantly evolving pattern of forest growth, starting with a field from which the forest has been cleared and abandoned.[1] The original forest that occupied the field will return, but only after a series of temporary plant communities have prepared the way. The successive stages of the new forest will each be different from that which ultimately develops. While subject to much discussion by ecologists today, Odum describes succession as being based on three parameters:

- it is the 'orderly' process of community changes which are 'predictable';
- it results from the modification of the physical environment by the community;
- it culminates in the establishment of as 'stable' an ecosystem as is biologically possible on the site in question. With each successive stage the insects, birds and animals change, giving way to other groups of creatures. Succession is rapid to begin with in early stages, and slows down in later ones. The diversity of animal and plant species tends to be the greatest in early stages and subsequently becomes stable or declines.

Succession starts with the invasion of grasses and other colonizing plants. Over time and depending on the degree of soil disturbance, these are replaced by shrubs and fast-growing pioneer tree species. These in turn are replaced by climax species, such as hemlock, oak, maple and beech. Once mature, a plant community is said to have reached a 'steady state'. The climax vegetation perpetuates and reproduces itself at the expense of other species that have been crowded out. Most ecologists consider this steady-state situation to be more of a theoretical concept than actual, however. Odum describes it in terms of bio-energetics; 'energy fixed tends to be balanced by the energy cost of maintenance'.[2]

Structure

Forest communities grow in a series of layers. The highest, forming the canopy, are the dominant species that control the environment for the rest of the forest. Beneath these are smaller understorey trees that are adapted to living in partial shade. At the lowest level on the forest floor are the ferns, mosses and herbaceous plants. Each basic group of plants is structured to take advantage of various conditions. The canopy species are exposed to maximum sunlight. They intercept much of the rain and evaporate a great deal of moisture back into the air. At the same time the canopy is an important modifying climatic agent for the life that exists beneath. The understorey plants growing in the shade of the canopy reproduce in the spring when it is bare and sunlight can reach the forest floor. At the level of the soil are the fungi, moulds, bacterial and other decomposers that recycle nutrients from rotting leaves and fallen trees and branches back into the system via the soil and roots. The number of layers varies depending on the forest type. Highly developed, northern deciduous forests usually consist of four strata. In southern communities the layers are more complex and some typical rain forests may be arranged in as many as twenty-seven groups.[3]

URBAN PROCESSES

Influences

As we have seen in Chapter 2, forests regulate the flow of water in streams and rivers and water storage underground maintains its purity and health. Forests have a great effect on the movement of water from the atmosphere to the earth and back again. They are thus interwoven with the physical and biological processes on which life forms depend. Cities have created modified environments to which natural plant communities have generally not had time to adapt. In terms of biological time, cities first appeared less than 10,000 years ago. Flowering plants are the product of an evolutionary process which began in the Mesozoic era some 200 million years ago. Trees have been exposed to more than 100 million years of selective pressures to adapt to natural environments.[4] However, their survival in the city is subject to many environmental pressures to which they have not previously been subjected.

The city climate is warmer, which has had a marked effect on plant distribution and survival. The atmosphere of industrial cities, as we shall see in Chapter 6, contains chemical pollutants such as sulphur dioxide from residential and industrial combustion, ozone from the photochemical breakdown of automobile exhausts, nitrogen oxides and fluorides and fine particles emitted from industrial processes. These pollutants interfere with the normal transpiration and respiration processes of plants. Their root systems, adapted to forest soil conditions, must cope with disturbed and compacted soils and paved surfaces. Such conditions reduce water penetration and supply of nutrients, lower groundwater levels and interfere with the transfer of air and gases. In northern regions, the soil is contaminated with salts that are applied to the streets of many cities to keep them clear of snow and ice. The attraction of trees for urban pets may contribute to the concentration of salts beyond their level of tolerance. There are also other physical problems that plants must contend with in the city environment: for instance, confined soils and exposure to cold from restricted or raised planting areas, exposure to heating and cooling vents on buildings, exposure to high winds or extreme heat, waterlogging from old basement structures or excessive dryness, the altered structure of urban soils, continuous disturbance from construction and maintenance activities.

Urban plant communities

The impact of the city has been far reaching on plant communities. There three general groups that deserve attention.

The cultivated plant group

This is the group of plants that is the product of horticultural science – the cultivation and selective breeding of plants that will satisfy the environmental and cultural demands of urban conditions. Through cloning and grafting techniques a range of plant species has been, and continues to be, developed in response to increasingly stringent requirements. For example, a large portion of the seed and stock of Canadian shade trees has been imported from the US and western Europe.[5] These must be resistant to a growing number of diseases, leaf rusts and insects. They must withstand drought, restricted soil conditions and doses of road salts. Their branches must be resistant to breakage from high winds or snow loads. Their leaves must tolerate poisonous gases and particulates in the urban atmosphere. They must also respond to rapidly changing mechanical constraints. For instance, roots cannot interfere with underground services, branching habit and height must not compete with overhead wires. In addition to a multitude of environmental and physical limitations imposed on trees, their selection and breeding has been dictated historically by prevailing aesthetic values and conventions. How should trees behave? What is their ideal form? How profusely can they be made to flower?

The ideal city tree must be fast growing, but long-lived. It must be symmetrical and perfectly formed. Characteristics such as messy fruit or slippery leaves, thorns, peeling bark and other inconveniences are unacceptable. These requirements are built into every designer's specifications for planting and have helped set the aesthetic and cultural standards for plant nurseries. The physical conditions under which plants are grown in nurseries are geared to uniform standards of soil, resistance to disease and form. They are grown to be transplanted to equally uniform sites. The unwary designer attempting to select moisture-loving plants for a wet, poorly drained site may well find they succumb to these conditions. The nursery plant, grown on well-drained loams, is no longer adapted to its original environment. The procedures necessary to ensure the survival of plants involve complex engineering and horticultural requirements for transplantation, moving, soils preparation, guying, irrigation systems and protection.

The native plant community

In many cities one may still find native plant communities that have remained relatively unaltered. These remnants of natural forests or wetland have been surrounded by the advancing city, but still retain elements of the original ecosystems that once prevailed. In some North American cities they survive more by good fortune than by design. Topographical obstructions or, more recently, planning policies, have forced the city to move around them. Some have since been incorporated into urban parks systems. The Tinicum Marsh and Fairmount Park in Philadelphia and the ravine lands in Toronto are examples. Others, such as the common lands of forest and heathland in British and European cities, have been kept open for public use since the first days of village settlement and have never been enclosed or cultivated.[6]

Many natural areas have been encroached on by transportation links and development. Changes in drainage patterns, erosion and human use have severely altered their original character. But in spite of these disturbances something of the natural diversity of the original natural community can still be found. Many wooded remnants still retain plants and animals that have become locally rare due to their isolation in the urban region. It is here in the middle of the city that one can still find trout lilies flourishing on the forest floor, or observe the annual migration of birds. Such places are one of the irreplaceable links between natural and urban processes. They are a small but vitally important historic and educational opportunity for nature in the city.

The naturalized urban plant community

These are the plants that have adapted to city conditions without human assistance. The evolution of civilization has fundamentally altered the original ecosystems that once flourished in the absence of humankind. Pollen analysis has shown that from its earliest beginnings the disturbance of land has been associated with plants of the open ground

Plate 3.1 *The natural progression to the 'ultimate tree' for the consumer society. A plastic lollipop stuck to a pole: no watering, no spraying, no pruning, permanent or throw away model*

Plate 3.2 *The ideal cultivated street tree*

Perfectly formed, even aged and resistant to disease. The selection process has provided the landscape design industry with a small group of plants bred to withstand any kind of environmental condition and usable everywhere regardless of place

Plate 3.3 *The native plant community*

Remnants of old forests that have somehow survived, even though they have, in many cases, been drastically altered. They provide the irreplaceable natural history and educational links with nature in cities in all seasons

(Photo: Steve Frost)

a

b

a

Plate 3.4 *Naturalized plants downtown*

*In many parts of the city, dense stands of trees, regenerating on their own, provide essential summer shade, where it is needed, to buildings and paved surfaces. The compound leaves of Tree of Heaven (*Ailanthus altissima*), a tree widespread in North American cities, sprout late and drop early, minimizing obstructions to the spring and winter sun*

a *A dense stand of ailanthus, crowded into a parking lot*

b *Plant diversity in a regenerating residential lot*

c *(opposite) Emerging through basement area ways, a testament to the vigour of nature in cities*

b

c

such as the dandelion, chickweed and plantain. These plants colonized the land following the Ice Age. Their habitat and distribution were subsequently restricted by the afforestation that followed, but expanded again when agriculture and cities again laid the soil bare.[7] Modern industrialized cities have had a profound influence on plant communities. Many have been lost or become extinct through the disappearance or changes to habitats in which they are able to grow. At the same time new associations have become established in humanized habitats as a result of species migrating from other climatic zones. While urbanization reduces the amount of vegetation, it has been shown that in European cities there is a comparatively high number of species present compared to the surrounding agricultural countryside.[8] The types of plant associations that have adopted the city are different from the native associations that the city replaced. Continual disturbance of the urban environment creates ecologically unstable conditions that are favourable to the invasion of numerous pioneer species typical of early succession stages of natural ecosystems. Ragweed, for example, migrated across Canada eastwards from the prairies where it originated as the railways, highways and settlements prepared the way.[9] The increasing number of alien species colonizing the European city has been shown to have originated from warmer areas of the world.[10] Their success is due to a warmer climatic environment; their expansion and naturalization have been greatly influenced, not only by direct introductions and environmental changes, but also by spontaneous hybridization with cultivated plants. From these, new species are continually evolving, adapted to the special soils and climatic conditions of the city.

The city offers a wide range of sites for these naturalized urban communities. They include waste lands that have been created by the demolition of old buildings, abandoned waterfronts and industrial lands, railway embankments, road rights-of-way and similar places. It is here that associations of white poplar, Manitoba maple, dandelion, chickweed, yarrow and other species grow in vigorous profusion. Even in the most paved-over parts of the city, mosses and grasses colonize rooftops and walls of buildings, dandelions, thistles and docks push their way through corners and cracks in the pavement. Ailanthus finds a foothold in the foundations of buildings and appears through basement gratings, where the warmth, moisture and alkalinity of the soil give it places to flourish. Plants, in fact, are everywhere in the city, a testament to their extraordinary tenacity and ability to evolve and adapt to the new conditions and environmental niches that the city provides.

Venice provides an excellent example of the phenomenon of urban plants in an intense urban setting that has, over time, become almost devoid of 'green' space in the conventional sense. Within the city limits 147 vascular plants have been recorded growing wild, including seven ferns and twenty grass species. The delicate maidenhair fern may be found in shaded cool corners of the city's footbridges, a rare plant originally from the warmer wetter coasts of Britain. Rock samphire grows along the edges of the canals, ivy-leaved toadflax on the Bridge of Sighs and many other plants take advantage of the special micro-climates created by the city's nooks and crannies.[11]

PERCEPTIONS AND CULTURAL VALUES

Horticultural science and the aesthetic that has evolved with it is ingrained in our urban tradition. The standard dictionary definition of horticulture – the art or science of cultivating or managing gardens – embodies the ideal of nature under control. Each tree, shrub and flower is a symbol of human ingenuity, an artifact in a humanized landscape. Like sculpture, plants are moulded and shaped, admired for their form, flower, leaves, unusual character or uniform repetition, as individual specimens not as part of a community. Set in an unrelenting carpet of mown turf, the ornamental trees and shrubs are as unchanging as the architectural setting in which they are placed. There is an unending human struggle to maintain order and control. It is evident in the formidable array of machines, fertilizers, herbicides and manpower, marshalled to maintain a

Plate 3.5 *Advertising with formal landscaping on the Gardiner Expressway, Toronto*

An unchanging setting that challenges the credibility of plants as living materials, and trivializes their fundamental role in sustaining life on earth

landscape as closed as possible to the form in which it was conceived. The dynamics of plant succession are subjugated in our expectations of how plants should perform and behave in the city. The urban landscape is a product of conflicting values. It expresses a deep-seated affinity with natural things. The spring bulbs displayed in every civic space clearly demonstrates this emotion. But these expressions of nature take place only on our own terms, subject to standards of order and tidiness imposed by official public values. The diverse community of plants that flourish in profusion in the adjoining abandoned lot, in every crack in the pavement and invade every well kept shrub border and lawn, represent, in the public mind, disorder, untidiness, neglect. The too frequent landscape improvement, intended to 'rehabilitate' a neglected area of the city, replaces the natural diversity of regenerating nature with the uniform and technology-dependent landscape of established design tradition. Few would question the value of the formal as part of the city's civic spaces. Street trees survive in the hostile habitat of downtown largely through the technical science of horticulture. The problem is not with science but with its application and the assumptions that go with it. The preoccupations of research in developing plants that are more and more resistant to air pollution is a case in point. The picture of ultimate technological success, where the air we breathe chokes us to death but our trees flourish unscathed, is one of misplaced priorities. This frame of reference perpetuates universal standards of design and upkeep of urban landscapes. It is concerned more with a scientific interest in plants as individual phenomena than with their place in the economy of nature. Urban design values have long been adrift, dictated by aesthetic conventions that lack a firm foundation in process and function. Such values work in opposition to the principles of ecological sustainability, and deny its fundamental principles of process, economy of means, diversity and social opportunity. The problem lies in the wasted energy such practices perpetuate, in a lack of a fundamental biological role for urban plants that go far beyond marginal amenity values, in the idea that parks should serve only a single purpose, and in the fact that plants with no economic or productive benefits are seen to have higher value than those that do.

SOME ALTERNATIVE VALUES AND OPPORTUNITIES

A functional framework

Our task here is to explore the alternative functional frameworks for urban plants that urban ecology provides. It is from this that a new aesthetic and environmental purpose can emerge, a process that has begun in many western cities. It involves two basic shifts in perceptual thinking. The first has to do with a frame of reference that embraces co-operation rather than confrontation with nature's processes. The second involves an approach to land that derives its inspiration from traditional farming and woodland management that has remained relatively unaffected by the industrialization of agriculture, or can be found in alternative 'organic' farming practices. The sensory appeal of these landscapes is derived from their functional role as working environments and arises from the integration of many ecologically sound objectives – protecting soils, producing timber, conserving water, natural features and wildlife and providing places for recreation. To the established principles of design and aesthetic use of plants in cities are added multi-functional objectives inherent in ecologically sound management. 'The New Forestry', an approach developed in Washington State that encompasses the entire forest ecosystem of plants, soils, animals and micro-organisms,[12] can provide the inspiration for managing city landscapes. We stand to gain from this approach in economy, educational values and overall benefits. A range of opportunities are made available for a richer, more diverse and more useful environment, and as the basis for an investment in the land. These are explored in more detail in the following sections of this chapter.

The naturalized plant community: fortuitous succession

One of the characteristics of modern cities is the considerable amount of land that lies idle or underused along transportation routes and service corridors, empty building lots, old industrial workings and waterfronts awaiting renewal. Many areas have been sealed from public use by ownership rights, or simply left unreclaimed. These places, once destroyed and subsequently abandoned or neglected, have, over time, evolved as naturalized urban plant communities. W.G. Teagle has described the conditions in the industrial Midlands of Britain, where over two and a half centuries of industrial activity has left an indelible mark on the region. The original native heath landscape was substituted for one of hills and valleys, created by coalmining and quarrying, which greatly influenced plant distribution. The proliferation of canals in the nineteenth century encouraged the spread of aquatic plants and invertebrates which previously had been unable to establish themselves in the fast-flowing streams that drained the surrounding uplands. The railways that followed played a significant role in the dispersal of plants such as the Oxford ragwort.[13] Their embankments and cuttings also provided a sanctuary for plants in the immediate surroundings that were being destroyed.

At the same time, the purposeful and fortuitous importation of exotic species throughout the world has occurred at considerable environmental cost. Purple loose-strife, a plant imported into North America, is seriously affecting the productivity of native wetlands; Norway maple (a cultivated tree universally used on city streets) and the Australian eucalyptus have impaired the quality of native woodlands. Strategies for rehabilitating derelict urban areas, however, also involve other problems that must be addressed. They have to do with industrial soil contamination, the cycling of toxic materials through the ecosystem and their impacts on the health of life systems and industrial heritage. These will be addressed in the next chapter.

Such issues notwithstanding, naturally reclaimed industrial landscapes often have an ecological, historic and topographic diversity that is far richer than the land reclamation and rehabilitation programmes have created. The 'green desert' monocultures of grass and trees are no match for the complex relationships of plants, soils, water, topography, micro-climate, wildlife and the complex industrial history that are found there. In many cities the sheer cost of effecting such engineered 'improvements' has saved many significant habitats and led to urban parks naturalization programmes across North America, Britain, Europe and Australia.

These initiatives, many originally born of economic necessity and subsequently inspired by public awareness of environmental issues, are the most telling argument for the thesis under discussion, that the diversity, vigour, beauty and wonder of natural process is available at little cost to enrich the city. It needs only to be recognized. Look behind the petrol station, the forgotten space used as a junkyard, the alley tucked away behind the city's main thoroughfares or the backs of many abandoned buildings in the poorer parts of town, and you will find magical places of dense forests and varied groundcovers that have appeared on their own. An examination of the city's residential areas will reveal connections between the types of plants found there and the relative prosperity of the various neighbourhoods. Observation of the city reveals that there is a marked tendency in well-to-do neighbourhoods for vegetation to have a greater variety of exotic species. The less wealthy 'down-at-heel' neighbourhoods have less cultivated vegetation overall, but have a much higher proportion of native or naturalized species. Many of these places rely on such trees as the Manitoba maple and the Tree of Heaven that have become established on their own. So-called 'weed trees' often provide the only shade over streets and sidewalks, creating, in many cases, an environmental quality that is not often matched by purposeful design. A comment by the botanist Herbert Sukopp is relevant here: 'Ecosystems which have developed in urban conditions may be the prevailing ecosystems of the future. Many of the most resistant plants in our industrial areas and in cities ... are non natives.'[14] As urban society is beginning to realize, we cannot afford to rely entirely on technology to create urban landscapes while ignoring the real nature of the places we live in.

SOME ALTERNATIVE STRATEGIES

General concerns

Since the 1980s the grass-roots environmental movement has clearly recognized that the ecological processes operating in the city form an indispensable basis for restoring the urban landscape. The interrelationships of climate, geology and geomorphology, water, soils, plants and animals, provide the fundamental ecological information on which environmental planning and management of land are based. At the broad level of spatial design the remnant natural and naturalized plant communities that occur all over the city require cataloguing and evaluation. Places that have habitat potential for plants and animals – patches of woodland or meadow, valleys and other linkages – should be integrated into the planning and social network of city places. They provide, among other things, alternative opportunities for diverse and rich recreational and educational experiences. A biological classification and management approach to open space with respect to its sensitivity to human intrusion is required for cities. Dorney has proposed a method for rating environmentally sensitive areas where sites are ranked into five classes of sensitivity or significance. These are derived from an assessment of such factors as the presence of rare or endangered plant or animal associations, the size of the area and species diversity.[15] Parallel examples are the International Biological Program, or

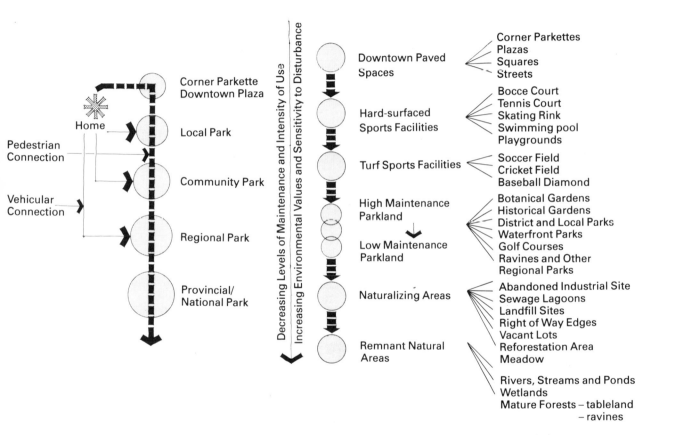

Figure 3.1 *An environmental classification system for urban spaces to supplement or replace the conventional hierarchy based on activities and size*

An ecological heirarchy that can classify spaces according to levels of sensitivity to human presence

Source: *M. Hough and Suzanne Barrett,* People and City Landscapes, *Toronto: Conservation Council of Ontario, 1987*

various non-urban parks classification systems. A 1987 system proposed by the author for classifying urban spaces, was based on ecological criteria and involved a sliding scale of open spaces, going from nature reserves of high sensitivity to human intrusion, and requiring restricted access, to downtown spaces that can support intense human activity and permit unrestricted access. In addition, a matrix was developed that relates different types of landscapes to various types of human activity, and provides a basis for deciding what activities should be permitted where.[16]

At a detailed level, ecological information on climate, soils, plant tolerance and succession creates opportunities for recognizing and acknowledging the uniqueness and variety of urban places. Design is based on the principle that solutions must be adapted to the site. Many alternatives become available once this framework is integrated into the planning process. The naturalized plant community becomes a valued resource for establishing vegetation on poor sites and sterile soils. Plants may be seen as constantly evolving communities rather than individual phenomena. Colonizing plants adapted to urban soils can enhance and modify them and provide alternatives to importing fertility to the city. Landscape maintenance can become a process of integrated management, based on ecological parameters, and gives us the practical tools for maintaining productive and self-sustaining landscapes.

Ecological restoration and plant succession: an alternative to the Green Lollipop

There is an aesthetic appeal to old trees rising clear from an open grassed sward uninterrupted by obstructions. It is a landscape that is achieved all at once, and is encouraged in urban parks because it permits maximum use of the ground plane for human activity. But this is a landscape without a future; the ultimate death of the old trees leaves nothing in their place. It will be years before the new trees planted now can recreate the original environment. The countless streets that were changed overnight by Dutch Elm disease are evidence of this fact. The propensity for creating static landscapes from climax species (what has been called the 'international urban tree community')[17] derives from an aesthetic perception of nature that sees beauty in decay rather than in the vitality of continuing life. This problem is admirably expressed by Guldemond:

> Our grandchildren will want to see old trees, so we will have to plant them. That means we cannot just wait and let the old tree stands die out.... We have to [consider] natural principles, which implies a normal distribution of age classes.... I always have a rather uncomfortable feeling when I see a park full of

Plate 3.6 *A wood without a future*

The classic woodland stereotype when translated into reality; nothing to replace the grand old trees

wonderful old trees, splendid to look at, but without any young material for the future and so without the permanence we all wish.[18]

As we saw earlier in this chapter, the forest environment performs many functions. From the point of view of human use, forests have throughout history provided the essential basis for the growth of cities. The care of forests is evolving towards a concept of management that has many objectives – economic, social and environmental. Harvesting of wood, recreation and amenity values, education, preservation and the maintenance of forest bio-diversity, all have their place in a holistic management process. The planning and upkeep of urban parks are, however, limited primarily to amenity and leisure objectives. While this may be relevant to the pedigreed parks and civic spaces, it is quite inadequate for the management of natural or naturalized forest areas. It is also incapable of fulfilling the multiple functions that plants must play in cities.

Another approach to planting that reinstates the natural process is needed in order to provide a self-perpetuating adaptive environment in areas where traditional horticultural approaches are inappropriate. This may be defined as the process of naturalization which brings an ecological view to the design and maintenance to the urban landscape. It involves the introduction of natural landscape elements into the city that include the re-establishment of woodlands through the reforestation of some lands, the creation of wetlands where hydrological conditions are appropriate, the development of meadow communities through modified turf management and the establishment of varied wildlife habitats. These, outlined in the restoration of the Don River on pp. 51–70, are inherently productive self-regulating communities achieved through ecologically sensitive management rather than total maintenance control. When this approach is placed in context with areas requiring higher levels of upkeep, it achieves benefits in environmental, social and aesthetic diversity and in overall economy in energy, materials and manpower.

Restoring urban woodland

Urban forestry, an increasingly well understood concept in Britain, Europe and North America, involves the transfer of ecologically sound forest management practice from the rural to the urban setting. Its objectives are based on the premise that forests, existing or introduced into cities, function to create low-cost and self-sustaining landscapes. Urban forestry requires a management philosophy that integrates aspects of horticulture with ecology and provides environmental social benefits. An integrated management policy of this kind may also produce direct economic benefits in the form of forest products.

Urban reforestation also involves land that has often not supported trees for a long period of time. Its intent is to create diverse plant associations that are in harmony with the nature of the site, i.e., with its soils, topography, climate and related environmental conditions. In addition, its long-term objective is to rehabilitate sites that have degenerated over time through soil compaction, removal of topsoil, reduction of productivity and nutrients. The philosophy of woodland establishment under discussion is based on the principle of natural succession speeded up and assisted by management. It follows three general phases:

- an initial planting of fast-growing, light-demanding pioneer species that quickly provide vegetative cover, ameliorate soil drainage, fix nitrogen and stimulate soil micro-organisms, and create favourable micro-climatic conditions for more long-lived species;
- an intermediate phase of plants that ultimately replace the pioneers;

- a climax phase of slow-growing, shade-tolerant species that are the long-lived plants.

In practice, the planting of these phases may be done all at one time, or introduced at intervals in the development of the woodland. (In the author's experience based on evaluation of research plots over eight to ten years, however, the latter practice has proven to be more successful than the former.)[19] The initial and final composition, character and uses of woodlands will be quite different as they evolve. They provide varying and useful places from the beginning. Initially they become socially useful and durable in a short time, something conventional climax-oriented planting does not do. Later the landscape will evolve to a different character, responding to changing uses and environmental conditions. The approach obviates the current problems that occur when whole sections of urban plant materials die from disease. It also maintains a dynamic sense of process and diversity that gives us a realistic view of nature and a valid basis for design with plants. Various examples of the approach have been put into practice.

The successional method for establishing urban woodland can also be applied on city streets. Practised in the Netherlands, China and other cities where environmental benefits outweigh purely aesthetic ones, this involves planting fast-growing trees in combination with slow-growing, climax ones. As the former grow larger they are pruned upward to allow the latter to develop. Over time the climax species take over. This management technique can be constantly repeated by replanting other species in gaps left by dying trees. This approach has the advantage of maintaining a permanent street cover. It involves, however, a change from established parks planting and maintenance regimes, to a successional approach that requires an alternative urban forestry practice, with implications for cost, personnel training and the availability of appropriate tree species.

A naturalization programme for an urban parks system

The parks, parkways and waterways that form a comprehensive open space network in Canada's capital city are a part of its long-range development plan initiated in 1950 by the French planner Greber. Both in its planning and design quality it compares favourably with the best urban tradition to be found in North America and Europe. Its overall image as a landscape of beautifully manicured lawns, well maintained shrub borders, banks of flowers and cultivated trees has, over the years, come to be regarded as the ideal of civic planning. The image represents a period of history when this kind of intensive maintenance was considered an appropriate and proper expression of national pride and commitment to good design. Today it poses serious questions for the National Capital Commission – the agency responsible for the implementation of the plan. How can a programme of development and ongoing maintenance be sustained in the face of escalating costs in energy, equipment and manpower? With over thirty landscape corridors comprising an area of some 2,748 hectares to be developed and 3,000 hectares of turf requiring weekly maintenance, the financial burden has become increasingly difficult to bear.[20] Are the accepted quality standards for landscape development and upkeep appropriate for all city lands? These standards, apart from being costly, have resulted in a large-scale landscape that provides little variety or diversity from one place to another – an image of monotony. When examined together, issues like this make a cogent argument for seeking other ways of developing and maintaining the city's landscape.

With these considerations in mind, in 1981 the National Capital Commission initiated a naturalization programme for its parkway corridors that includes the creation of meadow lands through a modified mowing regime and an experimental reforestation study programme designed to gain long-term knowledge of methods for establishing new woodlands. This radical departure from conventional practice on the part of a major public organization represents the beginnings of an alternative approach to the urban landscape that over time will become low-maintenance, economical and self-sustaining.

The project began with a series of test plots designed to evaluate specific planting techniques, management procedures and public acceptance of the idea. While in general terms the approach taken was based on managed succession, there were many questions that required answers and that dictated the test plot design, i.e.:

- the proportions of various species through the successional range of plants suited to the soils and climate of the region. Four relatively simple groupings of plants were selected that related to the well and poorly drained sites;
- the most effective types of site preparation techniques relative to cost factors, manpower, competition and speed of plant establishment;
- the best methods of ground treatment to control competing plants such as grasses and damage by rodents;
- the types of management required up to the establishment of woodland (canopy closure) and subsequent management of the evolving woodland (thinning of stands in relation to long-range objectives).

Test plots were initiated in 1983. Monitoring and evaluation of the results took place in 1987[21] to record what had been successful and what had not, and to determine the best approach to woodland establishment for future naturalization projects. This led to a number of observations and conclusions:

- an approach involving managed succession is by far the most effective way of initiating woodland;

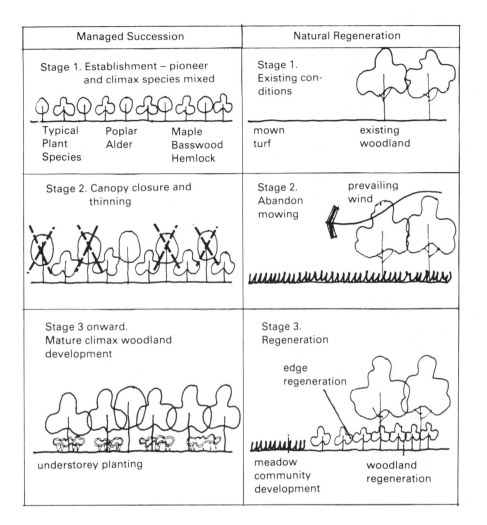

Managed Succession	Natural Regeneration
Stage 1. Establishment – pioneer and climax species mixed	Stage 1. Existing conditions
Typical Plant Species Poplar Alder Maple Basswood Hemlock	mown turf existing woodland
Stage 2. Canopy closure and thinning	Stage 2. Abandon mowing prevailing wind
Stage 3 onward. Mature climax woodland development	Stage 3. Regeneration edge regeneration
understorey planting	meadow community development woodland regeneration

Figure 3.2 *General reforestation categories*

Plantation involves the planting of predominantly similar species where the final woodland composition is determined by the initial planting. This is the normal procedure of forestry practice and is based primarily on commercial objectives.

Managed succession developed in the Netherlands and Britain is based on the principle of natural succession and assisted through management. The initial and final composition, character and uses of the woodland will be quite different as it evolves. The nurse crop functions to ameliorage soil drainage, fix nitrogen, stimulate soil micro-organisms and create a micro-climatic environment suited to the development of climax species. This approach is, therefore, concerned primarily with the rehabilitation of derelict landscapes, rather than with commercial objectives. Arguments on the advantages and disadvantages or native versus non-native plant species may be less important than considerations of structure, wildlife habitat, adaptability to soils, local climate, air pollution, drainage and so on.

Natural regeneration involves discontinuing mowing regimes in areas where a woodland seed source is available. In the absence of disturbance a woodland landscape is re-established naturally over time

Source: *Hough Stansbury Michalski Ltd, 'Naturalization Project', report, Ottawa: National Capital Commission, 1982*

Prescription	Planting Procedure	Comments

Alternative 1
Mechanical/manual cultivation
- cultivation of planting area prior to planting (fall) to kill ground vegetation
- manual cultivation regularly during growing season (monthly)

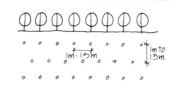

- labour-intensive
- application to small or awkwardly sheltered areas
- application to closely spaced planting where fast canopy closure is a high priority
- constant maintenance required during growing seasons

Alternative 2
Chemical treatment
- application of Round-up or equivalent herbicide to kill ground vegetation (fall)
- mechanical/manual cultivation of area 7-10 days following chemical application
- application of Simazine or equivalent herbicide (spring, prior to planting) in doughnut pattern around tree locations

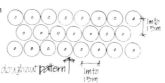

- greatly reduced labour requirements, since only one application required per year (depending on rate of application)
- applicable in small or awkwardly shaped areas and to closely spaced planting
- chemical treatment in urban areas may present problems of health and public acceptance

Note: Simazine is registered for use only for white pine and balsam fir plantations; while Simazine is also applied to hardwoods, many species are sensitive to chemical herbicides

Alternative 3
All mechanical cultivation
- mechanical cultivation of area prior to planting (fall) to kill ground vegetation
- mechanical cultivation between rows done regularly during growing season

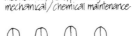

- greatly reduced labour requirements
- suitable for large areas where machinery may be economically used
- growing season cultivation still requires ongoing maintenance (monthly) until canopy closure

Alternative 4
- application of Round-up or equivalent herbicide to kill ground vegetation (fall)
- mechanical cultivation of area 7-10 days following chemical application, between rows
- application of Simazine or equivalent herbicide (spring, prior to planting) by mechanical spray between rows

- all mechanical treatment – therefore least manually intensive and least costly
- chemical treatment required only once a year (spring) depending on rate of application, until canopy closure
- requires large areas where machinery can be economically used
- chemical treatment in urban areas may present problems of health and public acceptance

Figure 3.3 *Site preparation alternatives*

Site preparation is a crucial factor in woodland establishment. It is necessary to reduce competition from herbaceous plants and rodent damage until the tree canopy closes and competing ground flora are naturally suppressed. The figure shows a number of alternative approaches under study by the National Capital Commission

Source: *Hough Stansbury Michalski Ltd, 'Naturalization Project', report, Ottawa: National Capital Commission, 1982*

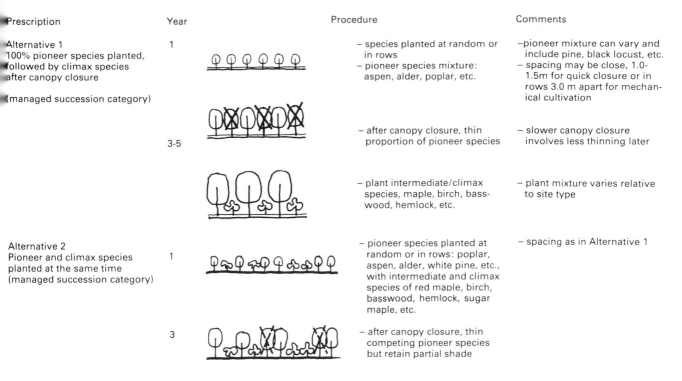

Prescription	Year	Procedure	Comments
Alternative 1 100% pioneer species planted, followed by climax species after canopy closure (managed succession category)	1	– species planted at random or in rows – pioneer species mixture: aspen, alder, poplar, etc.	–pioneer mixture can vary and include pine, black locust, etc. – spacing may be close, 1.0-1.5m for quick closure or in rows 3.0 m apart for mechanical cultivation
	3-5	– after canopy closure, thin proportion of pioneer species	– slower canopy closure involves less thinning later
		– plant intermediate/climax species, maple, birch, basswood, hemlock, etc.	– plant mixture varies relative to site type
Alternative 2 Pioneer and climax species planted at the same time (managed succession category)	1	– pioneer species planted at random or in rows: poplar, aspen, alder, white pine, etc., with intermediate and climax species of red maple, birch, basswood, hemlock, sugar maple, etc.	– spacing as in Alternative 1
	3	– after canopy closure, thin competing pioneer species but retain partial shade	

Figure 3.4 *Planting techniques*

The layout and spacing of plant materials depends on a number of interrelated factors that require investigation, for instance, the balance between closely spaced plants that achieve fast canopy closure but involve costly hand cultivation, versus widely spaced plants that achieve slower closure but involve cheaper mechanical cultivation; the relative merits of an initial 100 per cent pioneer planting versus mixing fast- and slow-growing species together

Source: *Hough Stansbury Michalski Ltd, 'Naturalization Project', report, Ottawa: National Capital Commission, 1982*

- poor soils are not an impediment to growth provided that reasonably well drained soils with a neutral pH can be established. Much of the soil in the test plot area was excavated subsoil from a nearby construction site, and was highly alkaline averaging a pH of 8. This required the addition of sand and sulphur integrated into the top 30 centimetres of soil;
- ground treatment that inhibits competition from grasses and winter damage to saplings by rodents can achieve a closed canopy of pioneer species and the beginnings of a forest floor association in four to five growing seasons, given the climatic conditions and leda clays of the Ottawa region;

a

b

Plate 3.7 *An example of growth rates in one plot in the National Capital Commission's research in woodland establishment*

a *View of plot 3 taken in spring 1985 at the time of establishment, showing perforated plastic and mulch treatment (background) versus tilled ground treatment (foreground)*

b *View of same plot four growing seasons later (autumn 1988)*

It shows the contrast between the low survival and growth rate of trees in the tilled portion of the plot (foreground) and the excellent tree growth in the plastic and mulch treatment portion (background). No maintenance was done on this plot during this time, barring watering during the first summer. Encouraged by long grass mice and voles decimated the foreground plantings

- of the various methods used to keep plots free from grasses and other competing species (manual weeding, rototilling, chemical applications, clover groundcovers), perforated plastic sheeting covered with mulch proved to be the most effective from the point of view of growth rates and management cost. In one plot prepared with this method, poplar pioneer species attained a height of 8 to 10 metres in four growing seasons;
- an initial close spacing of plants (about 1 metre) achieves a much greater survival rate and a faster canopy closure than the wider spacing used in conventional land-scaping.[22]

Using the techniques established in these experimental plots, a length of parkway was planted to native forest, and proved to be successful and economical as an alternative to the manicured approach that had traditionally been taken by the Commission's maintenance department. Figures 3.2, 3.3, 3.4 and Plates 3.7a and b illustrate some of the design and management principles on which the experiment was based and some of the results.[23]

The residential landscapes of Delft: the Gilles Estate

The landscape of many housing developments and parks in the Netherlands has been brought to a high level of sophistication and design. Developed in the late 1960s and early 1970s, the Delft Parks Service realized that a radically new approach to urban living was necessary, in contrast to the sterile and tidy environments of previous Dutch housing developments. The intention was to surround the housing with all the informality of the countryside and to create an educational landscape geared to unstructured, natural play environments. In the high-density apartment development of the Gilles Estate, court-yards were planted as urban woodlands in addition to providing open space for nursery schools, play areas and sports facilities. These inner court landscapes were based on certain fundamental and social objectives.

- There must be freedom of movement and play by children and adults. In these woodland landscapes the sheer vigour of early plant associations and their density provide a tough and highly varied environment. They must withstand the pressures of play and other activities, even in high-density housing environments. The Dutch believe that it is unrealistic to attempt to confine children to specific play areas, since this is not the way they behave in real life. While these are provided, the whole environment is available for play with a variety of opportunities and sensory stimulation for children ranging in age from pre-school to teenagers. The complexity and ruggedness of the woodland landscape accommodates pressures that would soon reduce conventional design layouts to ruins.

- Making a natural landscape. Plant material was selected on the basis of several criteria: to create intermediate scrub vegetation that would provide as rich a diversity as possible; and to emphasize the use of local native woodland species to minimize cost, maximize durability and for their ability to coppice. Plants were planted as whips in a random pattern spaced at 1 metre intervals. Specific site requirements dictated plant groupings, for example, locating willow in wet places, and the planting of larger trees to provide a mix of ages and initial shelter. Management responded to use patterns and circulation over time. For instance, secondary paths, created in response to natural desire lines, were gravelled and left as pedestrian and cycle routes once they had been established as permanent patterns of movement. As the woodland matured and responded to use, a self-perpetuating stable vegetation developed requiring little upkeep. The objectives of ongoing management were intended to produce a usable indigenous vegetation with a character of its own.[24]

To the observer, the quality of these housing landscapes are, therefore, radically different from conventional ones. They have the natural, informal, somewhat untidy, yet functional character of heavily used places. It is an aesthetic derived from the interaction of vegetation and human activity. A study of one apartment complex by the Institute of Preventive Medicine at the University of Leiden concluded that the landscape design and its use by children resulted in a greater attraction and more efficient use of the available space compared to the open spaces created by more conventional planning.[25]

A 1993 review of the Gilles Estate

The woodland landscape of the Gilles Estate was a revolutionary departure from convention for the Dutch which has had enormous influence in shaping the Green Cities movement in Europe and North America. The housing landscapes created by the Delft Department of Parks were a bold and imaginative response to the need to create more humane, varied and useful environments for the people living there than the previous housing developments of the post-war years. The three main post-war suburbs in Delft, of which the Gilles Estate was one, clearly show the changes in opinion as well as the financial constraints that shaped urban landscapes in the Netherlands.[26] Ton Jacobs, however, in his review of the Gilles Estate, suggests that the attempt to replace traditional horticulture by ecologically sound vegetation management has failed. The reasons have to do with the realities of human as well as natural evolution. He notes that while ecology is very much a

Plate 3.8 *The Gilles Housing Estate, Delft, Netherlands*

Comparison between the courtyard plantings in 1980 (a) and 1993 (b). Indiscriminate thinning in the interests of openness and safety has greatly altered the original character intended by the initiators of the concept

a

b

science dealing with change and development, it is also logical to consider human behaviour and their organizations in the same way.[27] The successful development of a plant community in an urban setting depends on the life cycles of the human environment. Jacobs describes the first Gilles residents as a homogeneous population, typical of the new suburbs, who moved into more spacious dwellings than they had been used to, and whose children loved the green landscapes of the estate. Over the years, children became teenagers and adolescents, parents got divorced and moved elsewhere, migrants from the Mediterranean moved in, policies changed to favour a wider spectrum of people, and the founders of the experiment left the Department of Parks to be replaced by people with less attachment to its ideals. Expectations and attitudes towards the green landscape, therefore, changed. Vandalism became an issue to be combated at all costs, an attitude that is not particularly favourable to green policies. And as the landscape matured people began to complain about safety,[28] demanding open well-lit spaces (an issue discussed on p. 125). The response was a thinning programme that removed much of dense understorey vegetation in many places. The long-lived species, however, were also cut and thus encouraged the continued growth of pioneer vegetation. Given the hierarchical structure of Dutch bureaucracy, the necessary changes in the relationship between park officials and the people from a 'them and us' attitude to a co-operative partnership did not mature, and, according to Allan Ruff, is still in an early stage of development.[29] In addition, environmental priorities among action groups focused more on issues with a more global impact such as acid rain and the ozone layer, with little concern for urban ecological issues. So as Jacobs comments, 'the Gilles experiment got international recognition as an experimental landscape. But local people hardly knew about it'.[30]

There are interesting comparisons between the perceived decline of this landscape and the LeBreton Flats experiment discussed on pp. 90–8. They have to do with behavioural and social management issues as well as the landscape itself, and how these must be integrated. In the light of this experience, a visit to Delft in the spring of 1993 by the author (just thirteen years after the first) was interesting for what it revealed about the place. Recognizing the validity of Jacobs' comments, there is much to be learned from what can be observed there. Much of the understorey has been cleared, but large patches remain, with woodland flora and understorey. There is an overall sense of tree canopy (even though much of it remains as fast-maturing pioneer species) and long vistas giving a sense of safety. There are areas of turf for group activities, and a wonderful quality of coolness. Children commandeer the place, using and playing in it as they did in the 1970s. Sunny private gardens for residents have been landscaped by them to their own tastes. Overall the place has a sense of ownership and control that is recognized by children and adults alike. The project, while considerably altered from the original intent, continues to evolve. Its sense of place is still powerful, and the lessons to be learned are still relevant to those who were originally inspired by it.

Issues of safety

Many communities with minimal or no experience of naturalized landscapes have questioned the viability of the Dutch approach when these landscapes are transferred to their own peculiar social and physical environments. Problems of safety and security consistently arise, which in many cases have the effect of restricting the potential to bring diversity and sensory richness to public places. The basic right of the public, particularly women, to feel safe from crime in their neighbourhoods, public streets, parks and workplaces is central to civilized life. Studies in the United States and in Britain point to the economic decay of downtown areas as a direct result of fear of crime.[31] The Safe City Committee of the City of Toronto has stated that sexual assault, the crime most often feared by women, is also a far more terrifying crime to contemplate than robbery, which is the crime most feared by men.[32] It is for this reason that a number of cities including London, Amsterdam and Toronto have focused on making public spaces safer through concentrating on violence against women.[33] The City of Toronto Safe Cities Committee has published a guide for planning and designing safer urban environments which includes ways in which the safety of places may be enhanced. The measures include:

- the need for appropriate lighting for pedestrians as well as for motorists in public spaces;
- sightlines – the ability to see what is ahead along a route, the avoidance of blind corners, impermeable landscape screens, etc.;
- avoidance of tunnels, pedestrian bridges, narrow passageways that offer no alternative choice for pedestrians;
- avoidance of 'entrapment' spots, such as small, confined areas, for example, recessed entrances;
- the need for places that have visual surveillance, where people can watch others;
- the value of activity generators, such as food stands that maintain informal surveillance of places;
- ensuring a sense of ownership or territoriality in neighbourhoods and public spaces;
- the need for appropriate signage and information about places, such as main routes, exit signs and main pedestrian routes.

These guidelines for safety and security are, among others, a necessary consideration in the planning for public space, depending on location, social environment and the specific characteristics of the place. They are, however, particularly relevant to the alternative strategies under discussion, since the relationship between naturalized landscapes in urban areas and safety is central to the realities of most modern cities that must be addressed.

Woodland parks and integrated management

The remnants of natural forest communities that have somehow survived in cities are irreplaceable, through often neglected natural elements. They stand as classic examples of the need for ecologically based management. Many have deteriorated from urban pressures, not the least of which is the colonization of remnant native woodland by exotic and cultivated species that leads to loss of species diversity, and general environmental degradation. The urbanization of watersheds alters stream erosion processes, sedimentation patterns, soil chemistry, vegetation and animal communities. The need to preserve both locally rare and common features, restore species diversity, protect wildlife and scenic quality is, therefore, crucial. Management objectives that integrate the physical and social influences of the surrounding city with the dynamics of a changed but evolving ecosystem under urban conditions provide irreplaceable opportunities for education.

Urban natural areas are the field study centres of the city, where plants and animals can be observed, where community dynamics can be studied and where the interactions of urban and natural processes may be measured. They are outdoor laboratories for teaching reforestation and silvicultural techniques. It is here that different management objectives must be integrated in the light of many and sometimes conflicting demands. They provide the essential links to understanding the nature and functions of forests beyond the city. Understanding the processes of nature and human intervention in familiar local surroundings is, perhaps, the most effective way of ensuring informed public concern for the larger environment.

In countries where significant rural timber is lacking, the city may play a role in producing timber to offset the costs of recreational facilities and park maintenance. Zurich, a city of a half million people, has a major proportion of its park space (nearly a quarter of the urban area) in forest and common land. These lands, some 2,200 hectares in extent, are within a half hour's street-car ride from the city centre and lie within or on the edges of the city. They have for many years been maintained on an integrated management basis, providing timber, recreation and athletic facilities, wildlife, agriculture, visual amenity and education. The forests are a mixture of deciduous beech, oak and maple stands and conifers. Forestry is carried out by the City Department of Forestry. It involves a variety of silvicultural techniques that include shelterwood, seed tree and patch cutting systems, depending on the forest type. The aim is to produce an unevenly aged forest of young and mature stands with a major emphasis on an aesthetic forest quality, and great care is taken in cutting, therefore, to ensure that this quality is maintained. Cutting occurs in the winter and the logs are stored along an extensive system of forest roads for shipping to various sawmills outside the city. Forest roads are also designed to be used for walking, nature trails, cross-country skiing and

Plate 3.9 *Zurich's forest parks*

Integrated urban forest management includes forest products, wildlife, small-scale farming and recreation. Sophisticated silviculture maintains the productivity of these woodlands and their aesthetic appeal. It shows how a park system can be, at least partially, economically and ecologically self-sustaining, contributing in ways other than recreation to the public good. As a multi-functional, self-sustaining landscape it provides social, economical and environmental benefits and calls in question the assumption that parks are exclusively for leisure

exercise by the citizens of Zurich. Fitness trails and exercise stops are also integrated into the forest setting and water for the park's picnic sites is supplied from the groundwater that the forest protects.

The basic aim of the forestry is to produce commercial timber for sawlogs and pulp. These products bring a return and help support the increasingly extensive and sophisticated recreational facilities that the city provides. In 1979 the balance sheet for the city's operations showed the costs of all recreation and forestry operations to be 4.5 million Swiss francs. This also included financial assistance to private clubs. Income from forest products amounted to nearly 2.5 million Swiss francs. Income, therefore, amounted to approximately 55 per cent of the total costs of the park system. An interesting aside on the economics of the operation is demonstrated in the sale of beechwood to Italy for fruit boxes. These arrive back from Italy with fruit, bought by the Swiss. The cost of the fruit is thus paid for by the sale of the boxes.[34]

It is apparent, therefore, that the concept of bringing rural occupations to the city in the form of urban forest products provides numerous benefits that conventional parks operations are unable to do. As we will see in Chapter 6, vegetation has a marked influence on climate and the urban environment generally. As well as the social benefits of a more varied and useful landscape, the urban forest can provide the basis for education in forest practice. Its presence as a part of the life of the city and under the scrutiny of its citizens ensures that high standards of forestry are maintained. It thus helps close the perceptual gap between urban and rural areas by creating a better understanding of their ultimate interdependence. In addition, it makes potential economic sense, a fact that has relevance in different ways for cities in both developing and developed countries that are trying to meet increasing demands for public amenities with diminishing budgets. Finally, the management of the Zurich forests, under full public scrutiny, makes the process visible and ties issues of forestry into daily life.

MANAGEMENT AND THE EVOLVING URBAN LANDSCAPE

Management and maintenance

The alternative ways of approaching plants that we have been examining offer design opportunities that provide a valid functional and ecological basis for urban form. They are also allied to the concept of continuous evolving management. The implementation of design, from paper planning to layout on the ground, is only the start of the design process. It presents a different picture, condition and usefulness at different stages, guided by a management process that determines its form over time. Conventional maintenance, however, deals with the landscape inorganically, as a static form. Its object is to keep the design as close as possible to the sketches the client accepted – the fixed picture. It brings with it a formidable arsenal of mowing equipment, leaf vacuums and blowers, fertilizer spreaders, herbicide and pesticide sprayers, to keep plants under control. The maintenance regime is high in energy and resources and aims to achieve standardized results. Obviously some types of urban landscape may require this kind of careful maintenance. Those subjected to intense human pressure, or those with intended gardenesque or historic objectives, are evident examples. But others do not. The role of management in cities must be to provide the greatest diversity possible, fitting a great many situations and needs.

The green carpet

Perhaps the most pervasive element of the cultivated landscape is the lawn. Instant wall-to-wall grass appears everywhere, in hot climates and cold, in prestige projects and in those that are run-of-the-mill, in large landscapes and small. Questions arise about the civic and political symbolism that 'official' landscaping represents. An article by Michael Pollan in the *New York Times* in May 1991 is indicative of the emerging values of urban nature that have arisen since the 1980s.[35] It suggested that if the President of the United States truly wished to be remembered as the 'environmental President' the White House lawn was not the appropriate symbol. It argued that

> [T]he democratic symbolism of the lawn may be appealing, but it carries an absurd and, today, insupportable environmental price tag. In our quest for the perfect lawn, we waste vast quantities of water and energy. . . . Acre for acre, the American lawn receives four times as much chemical pesticide as any US farmland.

The lawn is a symbol, in effect, for everything that is wrong with our relationship to the land, an expression of human control over a natural diversity that extends worldwide, from Britain to California to the Far East. According to the Lawn Institute in Tennessee, America has more than 80,000 square kilometres of lawn under cultivation, on which is spent US$30 billion a year.[36] The author continued with suggestions for more appropriate symbols of power and authority as a public setting for the White House grounds, such as a meadow that includes so called 'weed' species and a once a year mowing; a wetland, expressing one of the richest and most important of habitats; a vegetable garden, that could make the White House self-sufficient, or feed Washington's poor; and an apple orchard, productive, beautiful and *the* American fruit.[37]

The tradition of the lawn dates back to the nineteenth century and Frederick Law Olmsted. Olmsted was part of a generation of American landscape architects and reformers who set out to beautify the American landscape, although, as Pollan observes, 'That it needed beautification may seem surprising to us today.'[38] In 1870 Frank J. Scott published *The Art of Beautifying Suburban Home Grounds*. Intended to make Olmsted's ideas accessible to the middle class, the book probably did more than any other to determine the look of the suburban landscape in America.[39]

Today's turf depends on standardized requirements for topsoil, fertilizers, herbicides, watering and cutting heights, a total control that challenges its credibility as a living material. The lawn, maintained everywhere at 5 to 7.5 centimetres in height, is the product of the mowing machine which dictates the appearance and design form of much of the urban scene. Slopes greater than 33 per cent are difficult to mow. At less than 2

Plate 3.10 *Alternatives to mowing*

a *The fixed picture. Inorganic maintenance at work: hectares of grass at 5 centimetres, and maximum input for minimum output*

per cent they will not drain properly. Edges must be trimmed and defined. The mowing regime continues uninterrupted around and under other elements in the landscape, defying the variations of environment and moisture conditions that occur under trees, in and around groups of shrubs, on slopes and in depressions and in odd corners. Plants in the maintained lawn are thus restricted to the lollipop on a stick varieties, since the growth and spread by root suckering of many species is inhibited. A consistent high level of maintenance is self-defeating when universally applied and where it is not required. As a high-cost, high-energy floor covering, it produces the least diversity for the most effort. As a product of a pervasive cultural aesthetic in the context of public environmental consciousness, it defies logic.

There are simpler ways of producing a diverse landscape through a broadly based ecological approach to management processes. One alternative is a more intelligent and less intensive use of mowers to permit plant diversity and wildlife habitat to become management objectives. A study for the National Capital Commission in Ottawa on the implications of modified mowing regimes on the city's maintained grasslands showed the value of such an alternative. By cutting only those areas that were necessary for recreation, fire hazard and similar factors, and leaving remaining grasslands unmown

b *The creation of diverse meadow plant communities*

c *Cutting where it is needed, around groups of trees and shrubs, and associated with established community and road maintenance*

during the summer months, a far greater diversity of bird species was created in a very short space of time. The number of exotic 'nuisance' species such as starlings and house sparrows was reduced and the number of grasslands species such as bobolinks rose dramatically as the appropriate habitat became available.[40] This illustrates the need for grassland management objectives that take into account the recreational, visual and functional aspects of parkland that require areas of short turf, together with the encouragement of habitat diversity. If the maintenance of grassland rather than woodland regeneration is the objective, then cutting of unmown areas may be periodically required depending on objectives and the stability of the meadow. In order to preserve optimum grassland for bird habitat, mowing should be limited to late autumn, after breeding is over and after migrating birds have taken advantage of meadow seed sources.

Visually, the combination of rough and fine turf can add a complexity and variety to the landscape of a large grass area that universal mowing can never achieve. The problems that often arise when grass is left unmown are to a great extent perceptual. The image of neglect is nowhere more apparent than where the interface between mown and unmown turf is poorly considered. Turf left to grow long adjacent to human activities has tended to represent neglect and abandonment of responsibility. There are also practical aspects to the perceptual problem. Naturalized turf close to walks, roads or housing collects litter which involves higher maintenance costs to remove; and there is always the risk of fire. The creation of well-designed edges that accommodate mown areas for recreational and functional use while maintaining graceful lines, establish a sense of purpose and design intent. There are numerous opportunities for creating herbaceous groundcovers and these are used extensively in Dutch parks and open spaces. They are also being used to rehabilitate industrial spoil heaps in Britain and elsewhere. A survey of metalliferous mine wastes in Britain included field experiments which suggested that with the high concentration of toxic metals, rehabilitation could be more effectively achieved with naturally occurring grasses than with commercial varieties. Naturally occurring populations grew faster and persisted longer. They provided excellent stabilizing cover with the application of adequate fertilizer, and have persisted for many years. Three cultivars are commercially available that are tolerant of various pH and metal concentrations.[41]

The options available and their relevance to different situations are discussed in a report for the Nature Conservancy Council in Britain.[42] They include the introduction of native herbaceous flora and meadow communities. The conditions under which these thrive in the nutrient-deficient soils are typical of much derelict urban land. This in fact, is the opposite condition demanded by cultivated turf grasses. The elimination of imported topsoil and fertilizers makes their installation a great deal cheaper. Their floral diversity and low long-term maintenance are ideally suited to large-scale reclamation.

Plate 3.11

a *Orchard in the Netherlands: sheep maintain the turf groundcover reducing competition for the fruit trees, and providing another source of income for the farmer – a good example of economy of means*

b *An alternative to mowing behind fences: sheep in an oil tank farm in Greater Toronto. Here short turf is needed to prevent fire and maintain groundcover, and sheep do the job admirably for little or no cost*

They are applicable in such places as roadsides, vacant lots and similar areas where pressures of human activity are low. The high-resilient wild areas of Dutch landscapes have been based on high-nutrient input and acceptance of 'weeds'. These communities are seen as constantly evolving systems that adapt to wear and tear. This kind of ground surface is fortuitously created when turf along pedestrian routes is replaced with colonizing weeds, resistant to trampling and often covering the ground more effectively than the original turf.

Another alternative for managing many of the city's grasslands is to replace machinery with livestock. The uniform green carpet quality of the lawn is a product of modern technology, which could not exist before age of fossil fuels. Grass meadows were maintained by grazing animals. The eighteenth-century English landscape garden was conceived as an extension of the pastoral, agricultural landscape of a livestock economy and it used agricultural methods for the maintenance of its grounds. Sheep are remarkably effective as mowing machines. They graze close to the surface, nibbling at the grass shoots to produce a low, even turf that is both appropriate to many recreational uses and visually attractive. Sheep also encourage species diversity by selective grazing, something mowing machines cannot do. They require few physical facilities or personal attention, barring protection from dogs, access to water and shelter from the wind. Anyone who has visited the upland sheep country in the north of England will attest to their hardiness and ability to keep precipitous slopes in almost immaculate condition.

Sheep and other grazing animals are extensively used as an alternative to mechanical mowing in many country parks in Britain.[43] In some towns and villages, churchyards and meadows are still kept down by sheep as a result of arrangements with local farmers. However, this practice, which was once common in cities, has now all but disappeared. But its advantages as an economic and practical alternative to modern machinery are considerable, as the occasional example reveals. Laurie points out that many urban commons that are kept in a fairly natural state, are maintained at an extraordinarily low cost in relation to their size and the use made of them.[44] Some urban commons are still grazed today: Newcastle in England is an example. A petroleum company in Toronto has used sheep to mow the raised berms surrounding its fenced fuel storage tanks since the 1970s and has applied this technique to its holdings in other Ontario cities. Sheep are bought from the local stockyards in the spring and sold back in the autumn. Ten sheep maintain 1.2 to 1.6 hectares of steeply sloping grassed banks in a close-cropped and fertilized condition. The cost from the stockyards of each animal was Canadian $50 per head in 1977 dollars.[45] Even without taking into account the resale value of the animals, $500 per season for grass cutting is considerably less costly than mowing mechanically; nor do mowers fertilize the grass. In addition, the company has no need for equipment or other maintenance outlays barring an occasional bale of hay in very dry periods and a water supply.

There are various factors to be considered if such a practice is to be applied to the city. The major limitation is domestic dogs. Dogs worry sheep and can cause grave injury. So the need for fencing is important. In the absence of basic research into effective means of keeping dogs and sheep separated, or of enforceable laws to keep dogs under control, this strategy has practical applications for those urban open spaces that are fenced, or can be adequately supervised. Industrial areas, high-security installations, private or quasi-public lands and cemeteries are examples.

The re-establishment of rural grazing management adapted to urban areas is needed. For instance, adaptation of rotation grazing and paddocks where sheep progressively graze an area may be appropriate to large sites. Continuous grazing, where animals roam the whole area, may work best for smaller sites. Reverting back to mechanical mowing in some areas every second or third year may be required to minimize the build-up of parasites. There are also animal husbandry factors, such as shearing, dipping and vaccination for parasite control, that require arrangements with farmers and health authorities. The practical economies of using grazing animals on urban grasslands is illustrated by the fact that the University of Manitoba mowed its lawns with a herd of thirty sheep in the late 1940s and early 1950s. The reason for this was a tight maintenance budget and the presence of an agricultural college on campus. The sheep were guided by a herdsman throughout the campus area and controlled by movable fences. It is reported that this practice was terminated due to criticism by campus horticulturists, who wanted to see more flowers planted as well as a more attractive turf.[46] This example is particularly interesting since it points to the strength of prevailing aesthetic values at the time. It also points to the need, in the context of an environmentally aware society, for a return to economic and environmentally sound means of managing urban space that also provide educational opportunities. The British country parks use sheep, cows and roe deer, as much for the education and enjoyment for visitors as for keeping the grass under control. The isolated zoo-like environment of the demonstration city farm that some parks provide for deprived city children has less educational relevance than functioning farm management situations where these are part of the daily life of the city. The Newcastle commons previously mentioned, are continuously grazed by cattle through arrangements with local farmers. The commons are also in daily use for recreation and as a pedestrian link between different parts of the city. The presence of grazing animals is, therefore, an accepted part of urban life and establishes important links between city and countryside.

Management costs

The functional use of animals in appropriate places makes increasing economic sense the higher the cost of maintenance becomes. For instance, the maintenance of cemetery grounds, where upkeep is tied to perpetual care agreements, is becoming prohibitive and calls for practical, if unconventional, alternatives. The Dutch have made some interesting calculations of maintenance costs based on their woodland approach to urban design in comparison to more highly maintained traditional parks. Taking the maintenance of woodland landscape as the index against which other forms of maintenance are evaluated, the Parks Department at Rotterdam made the calculations shown in Table 3.1 for different types of plantings. From the table it is evident that woodland landscape is the least expensive form of planting. It is interesting to note that turf maintenance is over twice as costly, while annuals are sixty-eight times as costly as woodland. These figures relate to the Dutch experience and highly developed traditions in parks management. Also, many landscape developments incorporate a mixture of all or many of these categories. But they provide an interesting guide for comparison with other approaches, bearing in mind the unique problems that every city must face. They also have lessons for us in the development of low maintenance landscapes. Some comparative costs for woodland creation and naturalization illustrate the economies in establishment as well as continued management. It should also be borne in mind that with conventional landscape development these latter costs rise over time, whereas with naturalized landscapes they continue to go down.

Table 3.1 *Plants: comparative costs of maintenance*

Type of open space	No. of hours/100 m^2	Index
Woodland	1.7	100
Shrubs	1.8	118
Groundcovers	8.0	475
Roses	15.2	1,075
Rose gardens	18.8	1,291
Annuals	61.9	6,855
Perennials	29.9	1,842
Hedges	22.0	1,303
Lawns (overall maintenance)	3.3	210

Source: Aanemerij Plantsoerien vande Gemeete, Rotterdam, Department of Parks, Rotterdam, 1980

SOME DESIGN IMPLICATIONS FOR CITY LANDSCAPES

The discussion so far has dealt with alternative strategies that have included an ecological classification for city spaces, issues of naturalization and an ongoing management process that can be described as 'design over time'. We need now to examine city spaces as a whole, not only from an ecological and habitat perspective, but also from a social and behavioural point of view, since natural and human processes in the city are inseparably linked. The ability to choose between one place and another, each satisfying the immense social diversity of the city, has to do with quality of life. The tendency of parks planning to downplay these essential differences denies this inherent vitality and diversity. The basis for design, therefore, is multi-functional and multi-cultural, where compatible uses and site work together. The city's spaces have different potentials depending on such factors as their use, accessibility, biological and physical character, ownership, zoning, legal constraints, costs and practical applicability. Thus, as in the intelligent application of any integrated management plan, not all uses apply everywhere, or necessarily all at one time. But the application of this philosophy means that the traditional single use of urban spaces must be revised: i.e., recreation in the case of parks, sports and play areas, or motor traffic on residential streets. While different cities may have their own unique combinations of needs and resources, the basic outlines of the thesis under discussion are generally similar.

Residential parks and public spaces

Social and cultural realities and changing demographic trends are altering the nature of urban parks. The average uniform park for the average uniform white middle-class citizen is not a valid basis for an urban parks system. More innovative approaches to changing functions and spatial needs must be found for residential communities that parallel the ecological insights espoused in this book.

Streets

Possibly the most important function of the residential street is its role as community space. Yet an examination of space allocations in a typical residential area in Ottawa, Ontario, reveals the contradictions between planning regulations, jurisdictional liability and by-law and the realities of how the streets are used. With a density of about 71 houses per hectare the amount of space between house frontages taken up by streets is roughly 42 per cent of the total cross-sectional area. Not surprisingly, the streets are a focus of intensive social interaction and play by the adults and children of this

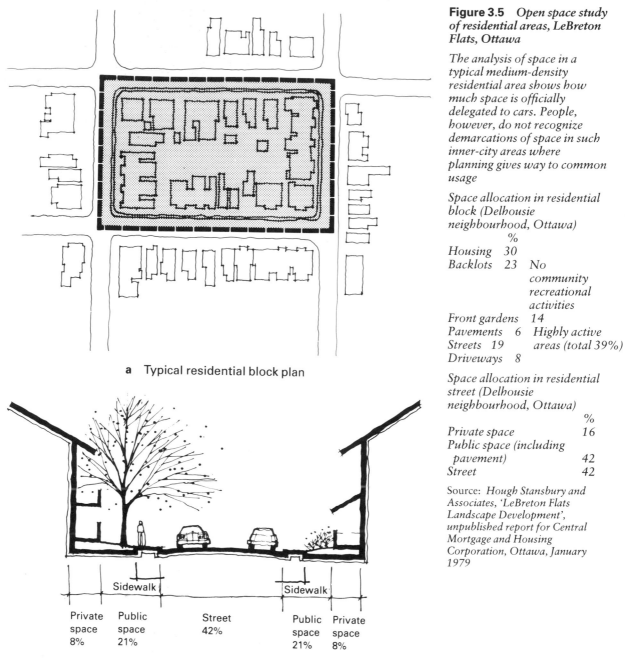

a Typical residential block plan

b Typical section through residential street

Figure 3.5 *Open space study of residential areas, LeBreton Flats, Ottawa*

The analysis of space in a typical medium-density residential area shows how much space is officially delegated to cars. People, however, do not recognize demarcations of space in such inner-city areas where planning gives way to common usage

Space allocation in residential block (Delhousie neighbourhood, Ottawa)

	%	
Housing	*30*	
Backlots	*23*	*No community recreational activities*
Front gardens	*14*	
Pavements	*6*	*Highly active*
Streets	*19*	*areas (total 39%)*
Driveways	*8*	

Space allocation in residential street (Delhousie neighbourhood, Ottawa)

	%
Private space	*16*
Public space (including pavement)	*42*
Street	*42*

Source: *Hough Stansbury and Associates, 'LeBreton Flats Landscape Development', unpublished report for Central Mortgage and Housing Corporation, Ottawa, January 1979*

Plate 3.12

a *The Dutch Woonerf. Designed as a multi-functional street, it is a social place for games, meeting, growing flowers or vegetables. Cars enter on sufferance*

b, c *A residential street in Toronto. Here common usage often supersedes the planning doctrine of separated uses*

French/Italian neighbourhood. The greatest social activity is occurring on space officially designated in the by-laws for vehicles alone. People do not recognize these demarcations of space and ignore them where they interfere with their normal patterns of living.[47] Planning by-laws gave way to common usage because the streets are the best places to do many things for which parks and other public spaces are less inappropriate.

The tendency to use streets as social space is spontaneous and natural as Jane Jacobs showed in the 1960s.[48] Since time immemorial the natural function of the street has been as a focus rather than a separator of social activity. A study of a residential area occupied by lower-income families in the inner city of Baltimore during a four-month summer period showed that half of the people counted, who could be determined as residents, were pursuing recreational activities. Only 3 per cent were using the parks; the remaining 54 per cent were in the streets, alleys, sidewalks and porches.[49] Mark Francis has suggested a number of ingredients necessary for the success of existing streets and for designing new ones.[50] Among these are:

- Use and user diversity. Healthy streets are used by different people for a variety of activities. Yet they are designed primarily for one group or a particular function, such as walking or driving. Lively and successful streets are those that recognize this need for variety. The social success of the 'Woonerf' that are widely used in Dutch cities is due to their multi-functional design, as places that foster a diversity of social activities, and where the car is not excluded, but enters on sufferance.
- Control. The sense of control that can be exerted over one's immediate environment is central to the social success of streets. Control is real for residents who maintain the sidewalk or street trees; it is symbolic when they feel that their private space, such as their front yard or entrance, extends into the public environment.[51] Control also has to do with such issues as safety and security, and with the environmental values of culturally different groups. Crime may be more controllable where a well established community has an investment in its own neighbourhood than one without that investment. Children and adults may well have different values. Francis points to a 1983 study of a new neighbourhood in Davis, California, that investigated what kind of playground should be designed for the children. The adults wanted a clean and safe play structure, while the children wanted opportunities for playing with dirt, water and natural elements.[52] This example also applies to the way groups with different cultural backgrounds respond to their own places, and the issue of who should design public spaces. Observation tell us that people with a long-term investment in their own places shape them in ways that are appropriate to their needs and tastes. While it has long been an article of faith that designers should be able to predict human behaviour the opposite is usually true, as we so often see in the playgrounds that children sensibly avoid, or the gathering spaces that are empty of

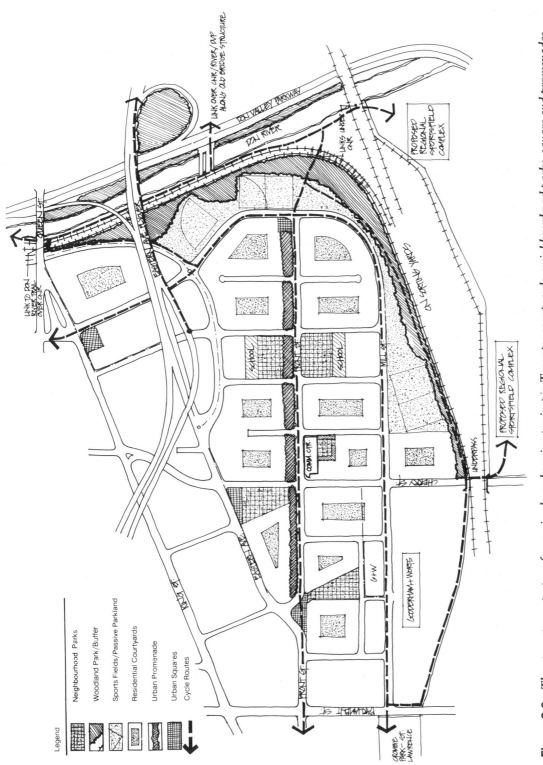

Figure 3.6 *The open space strategy for a mixed-use housing project in Toronto: courtyards, neighbourhood parks, square and promenades, sports fields and wooded buffers form a structure of places within which spatial control by people who live there can take place*

Source: *Hough Stansbury Woodland with Professor Alex Murray, 'Ataratiri Open Space Study', City of Toronto Housing Department, June 1990*

Legend

- Neighbourhood Parks
- Woodland Park/Buffer
- Sports Fields/Passive Parkland
- Residential Courtyards
- Urban Promenade
- Urban Squares
- Cycle Routes

Legend

potential: ● high ■ medium □ low/none

ACTIVITY/OPEN SPACE MATRIX

Possible Activities	"Natural" Area	Linear Way	Urban Plaza	Commu-nity Park	Sports Field	Sports Court	School Yard	Skating Rink	Urban pro-menade/ retail	Residen-tial Street	Totlot	Wading Pool	Court-yard	Rooftop	Balcony/ porch/ patio
Enjoy Nature/Education	●	●	□	■	□	□	●	□	□		□	□	■	□	□
Respite from city	●	●	■	■	□	□	■	□	□	□	□	□	■	●	●
Unstructured play	●	●	■	●	■	■	●	■	■	●	●	●	●	■	■
Walk	●	●	●	●	■	□	●	□	■	●	□	□	■	■	□
Bicycle/adult, teen	●	●	□	●	□	□	□	□	●	●	□	□	●	□	□
Bicycle/children	■	■	□	●	■	□	●	■	●	●	●	●	●	■	□
Jog	●	●	■	●	●	●	●	□	■	●	□	□	●	■	□
Walk a dog	●	●	■	●	●	□	□	□	■	●	□	□	●	●	●
Picnic/BBQ	■	□	□	●	●	●	■	□	□	□	●	●	●	●	●
Sunbathe	■	□	□	●	■	□	■	□	□	□	■	●	●	●	●
Socialize/hang-out	■	□	●	●	●	●	●	●	●	●	●	●	●	■	■
Be alone	■	●	□	■	□	□	□	□	●	□	□	□	□	●	●
Garden: veg/flowers	□	□	□	■	□	□	□	□	□	□	□	□	□	□	□
Toboggan	●	□	□	●	□	□	□	□	□	■	□	□	□	□	□
Skate	□	□	■	●	□	●	●	●	□	□	□	□	□	□	□
Tennis	□	□	□	■	□	●	□	■	□	□	□	●	■	■	●
Basketball	□	□	□	●	□	●	■	●	□	□	□	□	□	●	□
Baseball/soccer	□	□	□	□	●	□	■	□	□	□	□	□	■	■	□
Horseshoes	□	□	■	●	●	□	■	□	□	□	□	□	■	□	□
Skateboard	□	□	■	●	□	●	●	●	□	●	□	□	■	●	■
Street hockey	□	□	□	●	□	●	●	■	□	●	□	□	■	●	□
Festival	□	●	●	●	■	□	■	□	●	●	□	□	●	●	□
Outdoor theatre/concert	□	●	●	●	■	□	□	□	●	●	□	□	●	●	●
Window shopping	□	□	●	□	□	□	□	●	●	□	□	□	□	□	●
Workday lunch	■	■	●	●	□	■	●	●	●	■	●	□	●	●	●
Table games	□	□	●	●	□	□	■	●	□	●	□	□	●	□	□
People watching	□	□	●	●	●	●	■	●	●	●	□	●	●	●	■
Sidewalk performance	□	●	●	●	□	□	□	□	●	●	□	□	□	□	●
Car repairs	□	□	□	●	□	●	●	□	□	●	□	□	■	□	□
Frisbee/ball	■	■	●	●	□	■	●	■	□	■	■	□	□	□	●
Swing, slide, climb	□	□	■	●	□	□	■	■	■	●	●	●	●	●	●
Play tag	■	■	■	●	■	□	□	■	■	□	●	■	●	●	□
Play in water	□	□	□	●	□	□	□	□	□	■	●	●	●	●	■
Play in sand	□	□	□	●	□	□	□	□	□	■	●	□	●	●	●

Figure 3.7 Activity open space matrix showing possible relationships between activities and places

people. The task of designing public spaces has more to do with creating the physical and institutional frameworks within which people can shape their own environments in accordance with their own changing needs.

This kind of approach was developed in a proposed mixed use housing project in Toronto. In a behavioural sense, it involved the concept of control, where the residents of an area have the ability to regulate events and conditions in their neighbourhood.[53] It was recognized that this was based on the need to recognize, first, the psychological implications of territory – what an individual or a group sees as controllable space belonging to them, and what can be regarded as space belonging to the general public – and second, the tenure form of housing development that determines the degree and character of spatial control that residents are likely to exercise – i.e., co-operative, non-profit or commercial rental.[54] Conventionally, open space has been defined as either public or private. This, however, contributes little to behaviour-based design because it tells us only about who owns the space, not about how it will be used. To design places that will support and encourage individual and community development, it is necessary to know who owns it, who has access to it, who controls how it is used and who might go there.

The type of spatial control that emerged was based on the recognition that people using spaces would belong to one of two general groups: strangers, who neither live nor work in the development and neighbours who do live and work there.[55] The conceptual hierarchy or structure of spaces that was adopted included residential courtyards, neighbourhood parks, urban squares and promenades, sports fields and a major woodland buffer. It reflected social and physical diversity, a commitment to local neighbourhoods, the potential for long-term investment in the community and the opportunities presented in the site itself. An open space matrix was designed to help in the identification of spaces and their potential uses and activities. It illustrated how a parks system can support many different people with many different interests. Contrary to much typical open space planning that dictates the use of space, its intent was to provide a flexible framework within which decisions could be made by the residents themselves as the community matures.

Vacant lands

As we shall see in Chapter 5 (p. 203), there is a growing movement in many cities in the western world towards resident control and management of neighbourhood space that parallels an increasing difficulty, on the part of public authorities, to provide public amenities. Vacant land created by uncompleted urban renewal, the abandonment of city core areas or the economic decline of industrial areas, permeate the urban landscape and

Plate 3.13 *An increasing number of abandoned railway lines provide opportunities for linkages within parks systems and between cities*

Here a rail bridge forming part of the 1880s circuit around Toronto, and now a pedestrian way called the Iron Horse after the train that once used the line

(Sculptor: Robert Sprachman)

contribute to neighbourhood deterioration. Vacant patches of land between buildings where colonizing grasses or trees may be found, corners around parking lots and school yards as well as designated park spaces, provide important opportunities for creative play and environmental education that should be recognized. In some cities, residents have begun to assume responsibility for transforming unused land into productive open space for recreation, play and community gardens. The Federation of City Farms in England and Europe, the Trust for Public Land and San Francisco's League of Urban Gardeners are examples of grass-roots organizations that are concerned with these emerging social and environmental objectives and with practical ways of achieving them (Chapter 5, p. 203).

Communications links

Transmission lines, railway rights of way, canals and highways are uses that consume enormous quantities of land and provide considerable potential for urban space. These links, in particular abandoned railway lines, major utility rights of way and canals, have several characteristics that will determine the opportunities for alternative functions:

- they provide physical and potential biological links through the city to the surrounding countryside;
- for the most part they have little active use, being regarded in many cities as 'waste lands';
- public access is in many cases restricted for reasons of security or ownership.

Right of way lands are often colonized by naturalized plant associations that have succeeded on their own where there has been little or no disturbance for some time. Many harbour plant species which are not found elsewhere on lands subjected to horticultural management. These characteristics indicate that communications links have environmental and social value as corridors for the migration of plants, wildlife and people. Their planning is, therefore, associated with recreational access, education and as reserves, and management should reflect these values. Much highway land is taken up with interchanges that are often maintained as mown turf. Reforestation and naturalization of these areas, within the limits of traffic safety, reduces maintenance and provides improved wildlife corridors, as practice in North America, Britain and Europe clearly shows. The potential contribution to recreation uses and pedestrian linkages of urban highways is often overlooked. Many could make use of underused rights of way to provide pedestrian and cycle connections through the city, a common practice in some European cities that integrate pedestrians and vehicles as basic components of transportation planning. Left-over spaces that result from expressway interchanges could also become useful parts of the park system if connections were made to them. Links by

Plate 3.14 *A cycleway in Wageningen, the Netherlands, with a separate walkway for pedestrians*

Plate 3.15 *A nature trail attached to the Central Electricity Board's 400 KV Pelham substation, situated on the borders of Essex and Hertfordshire*

(Source: *Central Electricity Generating Board, London*)

means of underpasses or overhead connections could give such areas new recreational dimensions, as occurs in the urban expressways of Stockholm.

Railways have become, since their construction in the nineteenth and early twentieth centuries, rich in naturally regenerated habitat and are also significant natural corridors. While active lines are a safety hazard to pedestrians, their location in the city often makes them the best short cut between destinations. They could, in some circumstances, therefore incorporate essential pedestrian linkages. The decreasing importance of railways in some metropolitan areas and the increasing number of abandoned lines provide opportunities for cycle and pedestrian access to link residential communities, parks, schools and commercial areas. The growing North American parks movement of 'rails to trails' is an example of where walking and cycling trails are establishing links between cities. The Rails to Trails Conservancy movement in the United States has been helping to bring about track conversions since the 1970s, and 6,436 kilometres of tracks are reported to have been converted to bike use during this period.[56] These greenways provide a network of trails along abandoned railway rights of way throughout the country. One of the first conversions was the 1967 Elroy-Sparta State Park Trail that crosses 51.5 kilometres of dairy country in south-west Wisconsin.[57] The considerable land holdings of railway companies that have lain idle for many years provide significant social and productive value. The Camley Street natural park created on abandoned land in the centre of London is a case in point.

Other linkages include the canal systems of nineteenth-century industrialization and electrical transmission rights of way that are increasingly recognized for their potential as connectors, both through and between cities. Electricity rights of way may pose height limitations for reforestation and are frequently interrupted by roads and other obstructions which limit their value as wildlife corridors. At the same time they are potentially a major provider of education and recreation space, often support a diverse group of pioneer plants and are valuable as wildlife habitats. In 1967 the Central Electricity Generating Board in the Midlands region of England opened its first nature trail, field study centre and nature reserve on a substation site, and since then others have followed, developed in association with county councils, education authorities and naturalist societies.[58] These corridors also provide much needed park space, market and allotment gardens, particularly where they are associated with residential areas. At the same time there is considerable debate and controversy over the effects of magnetic fields on human health. Some research has suggested that power lines pose risks of cancer, while others have found no such evidence. While there may be no clear-cut conclusions on this issue, communications links of all kinds have wide value for forestry, wildlife habitat, recreation and water-based connections when they can be integrated into the corridor planning network of the city.

Industrial lands

Industrial lands account for a large proportion of metropolitan areas, with large parts of them serving little productive use. A number of characteristics that are of importance to alternative uses are pertinent here.

- Many industrial plants are high-security operations and are fenced and inaccessible to the general public. Industrial buildings are often surrounded by large areas of closely manicured turf (particularly those that represent 'flagship' monuments to corporate aesthetic taste). Their potential for ecologically appropriate alternatives reflecting another set of environmentally based values needs to become common practice. In some situations this could be associated with various types of livestock such as sheep or geese, providing returns in maintenance, visual amenity and public awareness of rural occupations. Oil storage installations and municipal sewage and water filtration plants that have security fences are examples of these high-security operations (see pp. 193–8).

- The heat generated from many industrial operations has potential for greenhouse market gardening, linked directly with industrial plants, or indirectly on adjacent land. In some places the integration of industry with agriculture has potential economies as an alternative approach to maintaining agricultural soils within the expanding industrial edge of the city.

- Much land is unused or abandoned, often near waterfronts or in inner-city areas. Fortuitous colonization often combined with poor drainage has, in many cases, created areas of special botanical, wildlife and heritage interest. These are also often naturally protected from intrusion by security fences. Redevelopment usually ignores the rich natural heritage that it replaces, and reclamation often replaces natural diversity with 'green desert' recreational developments and aesthetic amenities. Many of these areas have the greatest value left as they are. Alternatively, some of their natural assets could be incorporated into new development. Design policies should recognize the inherent opportunities they represent to enrich the city's landscape and provide alternative places for the study of natural processes and history.

Industrial heritage and contaminated soils

The legacy of the industrial era is a pressing issue faced by all western cities. The great industrial environments of work – the ships, railways, grain silos and industrial steel mills – represent both an opportunity to celebrate past heritage, and a liability in terms of the contamination of soil and water they have left behind. More often than not, efforts to remediate polluted soils leads to the destruction of heritage.

The complex problems of soil and groundwater contamination have to be addressed at many levels – biological and technical, jurisdictional, financial and legal. The legacy of hazardous chemicals, heavy metals and floating hydrocarbons at groundwater level, involves the resolution of many issues, such as appropriate clean-up technologies, acceptable levels of site mediation, liability, appropriate policies and regulations. There are also issues of human health and the impacts of contamination on natural systems. For instance, the transfer or pathways of contamination into the food chain through plant uptake which is then passed on to wildlife is not well understood. Approaches to remediation vary in response to site, types of industrial pollution and environmental, legal, jurisdictional and political agendas; the need to develop an integrated ecosystem-based, rather than a piecemeal, approach; the need for public involvement. Methods for dealing with these problems include leaving the material where it is (such as contaminated river sediments), excavation and disposal (shifting the problem from one place to

Plate 3.16 *As the ecological restoration of the Emscher River in Germany takes place, part of the original channelized river system has been left as an historical reminder of the past*

another at considerable expense), treatment on site, where the soil being restored is left in place, and biological systems (relevant where contaminants can be broken down organically through micro-organisms). In Tacoma, Washington, strong environmental legislation can impose heavy fines for failing to clean up a site, and casts a wide net for liability that includes 'current owners and tenants without regard to fault, transporters of contaminated wastes, and lenders'.[59]

A major concern is the frequent conflict of objectives that occurs in urban renewal between rehabilitating contaminated soils while protecting industrial heritage and the complex natural succession that one finds in many old industrial areas. The case-study that follows describes a remarkable example of commitment to these objectives in the now decommissioned Thyssen steel works at Meiderich, in the depressed industrial region of the Ruhr Valley in Germany.

Urban renewal in the Ruhr Valley: combining ecology, heritage and economy

Strategies for the economic renewal of old industrial areas are the subject of intense discussion in western and eastern Europe, the UK, North America and Japan. In

Plate 3.17 *The decommissioned Thyssen steel works at Meiderich, Ruhr Valley, Germany*

A heritage site and leisure attraction that is already providing habitat for many species of plants and animals, some of which are regionally endangered

(Photo: Manfred Vollmer)

Germany, the need for major restructuring has become most obvious in regions where once prosperous industries such as coal and steel have become obsolete, leaving behind a ravaged landscape, large tracts of derelict land, massive subsidence from past coalmining activities, channelized rivers that carry both wastewater and rainwater, high unemployment and a depressed economy. The Emscher region of the Ruhr, an area of approximately 800 square kilometres with a population of 2 million, is a case in point where the primary goal of *ecological* renewal, has been driven by the necessities of *economic* renewal.[60] In 1988 a decision was made to initiate economic and environmental renewal in this area and an 'International Building Exhibition' (an implementation instrument with a long tradition in German architectural circles) was set up to initiate this process. An 'Emscher Park' IBA was then created by the North-Rhine Westphalia government to develop restoration projects. These include the ecological rebuilding of the Emscher River watershed involving the creation of a network of parks where channelized streams can be renaturalized, and where bio-diversity can be re-established, the building of six new decentralized sewage treatment plants throughout the watershed, the construction of some 320 kilometres of underground sewers, development of stormwater infiltration techniques and the separation of rainwater from wastewater, residential and industrial developments, new social, cultural and sports activities, and the protection of industrial heritage. With a duration of ten years, the purpose of the IBA is to bring together specialists and politicians to act as catalysts for change, promote high-quality architecture and city planning and provide a forum in which local authorities, private companies and citizens can develop projects. Many of these are financed jointly by the state, the cities and private enterprise, while others, such as the Emscher parks, are funded exclusively through public funds. Financial investment for the Emscher Park project over the next ten years is expected to be some DM 3 billion, coming in equal measure from public and private sources.[61]

Other interesting economic strategies are being used to achieve economic objectives. With no water quality regulations, the German government charges for pollution being discharged into the Rhine. Different pollutants are carefully monitored and incremental charges are levied accordingly. In 1993 the total cost of pollution to be paid by the municipalities was DM 20 million. By 1996, this will increase to DM 100 million, the money going towards providing infrastructure. Thus building sewage treatment plants becomes a priority, since by the end of the twentieth century, when the goal of comprehensive water treatment is to be reached, the cost of levies will become prohibitive.[62]

Protecting the decommissioned Thyssen steel works at Meiderich posed special problems. It became apparent that to dismantle such a massive installation and clean up the soils underlying the plant would involve enormous costs. It was less expensive to keep the steel works intact and make them safe for public use as a heritage site and leisure attraction for an initial period of twenty years. This extraordinary monument to the heavy industry of the past is, therefore, being preserved together with an emerging forest community that has grown up around and within it. Today this site provides a habitat for many species of plants and animals, some of which are included in the 'Red List' of endangered species in North-Rhine Westphalia.[63] To see and experience the new forests emerging within and around the industrial structures, with wildlife, butterflies and birds making the place their home, gives one a distinct image of ancient Hindu temple ruins emerging from the jungle.

Cemeteries

Cemeteries are often among the most valuable open spaces in cities. For some they may be regarded as a forbidden yet challenging playground for small children in search of conkers and adventure among the tombstones. For others they are places of quiet and repose, away from the noise of the city. Old established cemeteries, with their narrow roads, varied topography and vegetation, provide a secluded haven for walking, jogging, nature study and meditation. For many people old tombstones provide the best and most interesting records of local history.

The long-term cost of maintenance also poses problems that could be turned to advantage in these often pastoral urban landscapes. Some cemeteries have considered the use of sheep and geese to keep the grass mown,[64] which would, in addition, introduce a productive use to these lands. Situations such as this have occurred in the past, as we shall see in Chapter 5 on urban farming, and they still occur today in church cemeteries and graveyards in some English rural towns. The role of the cemetery in the conservation of wildlife habitats is also significant, since they enjoy seclusion from intense human activity and often provide conducive environments for animals and birds.

Native and fortuitous natural habitats

Wetlands

Natural wetlands and open bodies of water contain rich associations of wetland species and are susceptible to damage through groundwater depletion and water pollution.

Plate 3.18 *(opposite) Rock-climbing practice is one of the recreational activities at the Thyssen steel works*
(Photo: Manfred Vollmer)

Plate 3.19 *Cemeteries are often among the most valuable spaces in the city, for local history, quiet and repose and as permanently protected places*

Abandoned industrial sites may also support rich communities that have occurred through a combination of water impoundment and natural succession. Being enclosed or off limits, they survive as precious natural reserves. These sites are often durable features in the city landscape, but being in private hands are prone to demolition. As human-created marshes, sewage lagoons are rich in nutrients and support diverse associations of plants, invertebrates and birds.

Woodlands

Remnant natural woodlands often contain locally rare trees, shrubs and ground vegetation and are important habitats for a variety of animals and birds. Naturally occurring regenerating woodlands in such places as vacant lots, lanes and old residential streets, are tough and resilient and provide climatic and social benefits to less favoured parts of the city.

Ravines and valley lands

Those that contain steep banks and remnant native vegetation that are highly susceptible to erosion and damage or destruction from urban land uses, often contain highly diverse groupings of plants. They are significant places for passive recreation, trails and education. Valley floors are often disturbed by the removal of the original vegetation due to overuse or inappropriate activities.

Regeneration landscapes

Many landscapes have completely changed ecologically, hydrologically and topographically due to mining or similar types of disturbance. These regenerating landscapes can often sustain high pressures of use from children and adults. As changed but none the less vital naturally regenerating environments, they are of priceless educational and ecological value. Former wastelands recolonized by early succession plants are ecologically diverse, resilient and of considerable educational and social benefit in residential areas. Their preservation and inclusion into new developments would do much to enrich these places.

Regional parks and spaces

At a regional scale the open spaces that include river valley systems and ravine lands, private and publicly held lands such as agriculture and conservation areas, and those that are currently dedicated in one form or another to recreational activities, lend themselves to multiple functions. It is here that the parks system must be understood within its larger bio-physical context. The concept of linked parks, dating back to Olmsted's Greenways, and Phil Lewis' environmental corridors in the American mid-west, have become accepted planning strategies for the larger urban landscape. As William Whyte observed in the 1970s:

> Per acre, linear strips are probably the most efficient form of open space . . . when they are laid out along the routes people walk or travel . . . the spaces provide the maximum visual impact and the maximum physical access. The linear concept . . . provides us a way of securing the highly usable spaces in urban areas where land is hard to come by, and in time, a way of linking these spaces together.[65]

The layout of capital cities in the decades after the Second World War followed the parkway and greenbelt principles of the time. The Amsterdam Development Plan begun

Figure 3.8 *Some general categories of habitat type in urban open areas: wetlands*

a *Wetlands and open bodies of water contain rich associations of wetland species and are susceptible to damage through groundwater depletion and water pollution*

b *Abandoned industrial sites may also support rich communities that have occurred through a combination of water impoundment and natural succession. Being enclosed or off limits, they survive as precious natural reserves. Abandoned industrial sites are often durable features in the city landscape*

c *As human-made marshes, sewage lagoons are rich in nutrients and support diverse associations of plants, invertebrates and birds*

in 1935 proposed a network of greenbelts and recreation routes linking parks in the various neighbourhoods, districts and suburbs of the city.[66] The greenbelt plans for London and Stockholm were conceived as strategies for limiting urban expansion outward while permitting growth to occur as satellite towns beyond city limits. Similarly, the greenbelt surrounding Ottawa, together with its linear parkways that provided links to the countryside, formed the 1949 structure for that city, its 16,200 hectares of public lands being acquired during the 1960s. In social terms, therefore, a linked network of parks become more useful than isolated ones. This seems to be borne out from a climatic perspective as we shall see in Chapter 6. But the significance of the linked system in the 1990s also lies in its role in protecting and maintaining ecological diversity, as recent work in landscape ecology has shown.[67]

In 1991, an ecological study of the Ottawa greenbelt system was undertaken for the National Capital Commission.[68] The study's purpose was to examine the bio-physical functions of the greenbelt and surrounding environs, and to illustrate how an ecologically based vision would help guide future urban growth and planning. Of particular significance was the very different philosophical approach that it took to the greenbelt in comparison to that used in the 1950s. The principles on which ecological planning are

a *Ravines and valley lands*

Figure 3.9 *Some general categories of habitat type in urban open space: woodlands*

Those that contain steep banks and remnant native vegetation that are highly susceptible to erosion and destruction, often contain highly diverse groupings of plants. Valley floors in many ravines are often disturbed by the removal of the original vegetation for reasons of access, park development and so on. Many of these are resilient and can withstand considerable use

b *Remnant mature woodlands often contain locally rare trees, shrubs and ground vegetation as well as birds and other animals*

c *Naturally occurring regenerating woodlands, for instance, on vacant lots, lanes, etc., are tough and resilient and provide climatic and social benefits to less favoured parts of the city*

a *Many landscapes have been completely changed ecologically, hydrologically and topographically due to mining or similar operations. These regenerating landscapes can often sustain high pressures of use from children and adults. As changed but none the less vital natural environments they are of priceless educational and ecological value*

Figure 3.10 *Some general categories of habitat type in urban open spaces: regenerating landscapes*

b *Former wastelands recolonized by early succession plants are ecologically diverse, resilient and of considerable educational and social benefit in residential areas. Their preservation and inclusion into new developments would do much to enrich these places*

c *Power line rights of way are the connecting links through the city, support a diverse group of pioneer plants and are valuable wildlife habitats*

Figure 3.11 *Regional parks and open spaces*

Linked small parks for social use and maintenance of larger areas for conservation

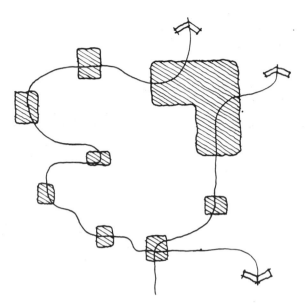

based involve an understanding of landscape patterns that are derived from life systems, and which can be identified and mapped in the planning process. In this sense, the biological and physical patterns of the land present a more deterministic and compelling basis for structuring and defining the shape of urban growth that the predetermined 'lines on a map' of previous planning theory.

The study examined the natural features that are contained within and outside the greenbelt boundaries, and the degree to which its boundaries protect them. The Ottawa/Carlton urban region as a whole has a highly diverse and extensive body of natural areas. They maintain an extensive genetic pool, and serve as headwaters for numerous rivers and tributaries. At the same time development pressures (including transportation corridors, servicing and infrastructure, recreation and building), advancing from outside the greenbelt as well as from within, have, to some degree, adversely affected all of them. Of particular interest was the fact that the majority of significant natural areas in the region were either unprotected, or at best only partially so, by existing greenbelt boundaries: a fact that clearly demonstrates the need for a new approach to greenbelt planning; one that is driven by regional ecological design.

In order to better integrate human settlement patterns with functioning natural

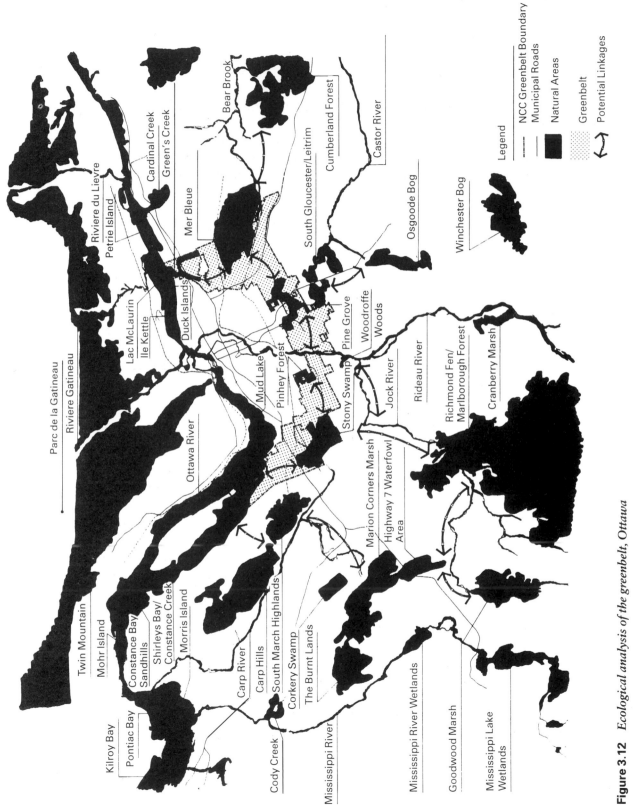

Figure 3.12 *Ecological analysis of the greenbelt, Ottawa*

A regional view of natural features shows clearly that the greenbelt does little to protect the natural systems on which much of the environmental quality of the city depends

Source: *Hough Stansbury Woodland/Gore and Storrie Ltd, 'Ecological Analysis of the Greenbelt', Ottawa: National Capital Commission, 1991*

systems, the recommended plan was based on three essential components:

- areas of high biological significance within which development is prohibited and recreational use restricted;
- buffers that protect primary areas from incompatible land uses; and
- linkages which connect primary natural areas and act as movement corridors for flora, fauna, water and people. These were given levels of priority relative to their biological significance, from a first priority that maintains essential links between significant functioning natural systems, to a third priority that includes major recreational corridors or greenways that also provide habitat linkage.

The land use policies formulated for the ecological plan for the greenbelt were consistent with the intention of the Federal Land Use Plan 'to reflect the special functions of the Capital so as to enhance the Capital's image'. One of the underlying policy principles of the Federal Land Use Plan notes that the Capital should 'display its key institutions, facilities, and symbols on prominent visible sites ... [to] ensure a built environment that is in harmony with the natural setting'.[69] The significance of regional parks and the greenways and greenbelts that link them, therefore, lies in the application of ecological principles to developing a landscape structure that will guide future urban growth, and where the interdependence of human and non-human systems is a primary goal.

In this chapter we have examined the natural systems of the city from the perspective of its native, fortuitous and cultivated plant associations. It becomes clear that the conventions that have created the cultivated landscape have been based on technological imperatives that are in confrontation with the realities of urban nature. They relentlessly overpower even the simplest observation of natural process at work from which much can be learned. We have the anomaly of a design process dedicated to urban quality, but instead creating impoverished environments. Naturally diverse places are replaced by a landscape of turf and cultivated plants that minimizes ecological diversity and social options. The alternatives available to us are rooted in an ecological view of plants as communities; a holistic management framework drawn from environmentally sustainable forestry and agricultural traditions; and an increasingly aware public that is initiating countless naturalization projects in cities worldwide. The concepts of urban forests, planting design founded on succession, grassland management and the larger structure of city spaces that bring together natural process and human behaviour, provide benefits in a more diverse environment, greater economic and environmental productivity and greater social and educational values. These are highly relevant in various ways, not only to industrialized countries but also to developing ones. In addition, the economies over current landscape practice are undeniable, and are increasingly becoming part of an emerging set of environmental values.

It is also apparent from our discussion up to this point that water and plants are indivisible parts of the natural processes of cities that must be seen as a whole. In this way they provide us with a useful and enriching urban environment and a means of closing the perceptual gap between the city and the larger non-urban landscape. A framework for an appropriate design aesthetic becomes available; one that is responsive to ecological, economic and social criteria. It is abundantly clear that plants are a fundamental part of the urban scene. Their significance as habitats for wildlife, for people and for ecological health varies in proportion to the complexity and variety of their associations. Thus the ideas discussed in this chapter have a direct relationship to the diversity and stability of all species of wildlife, and thus on the quality of the human environment that is the city. Since its health must now be gauged by the sum of its life processes, we must turn to an examination of wildlife.

4

WILDLIFE

•

The Dodo used to walk around,
and take the sun and air.
The sun yet warms his native ground –
the Dodo is not there!
The voice that used to squawk and squeak
is now for ever dumb –
Yet may you see his bones and beak
all in the Mu-se-um.[1]

INTRODUCTION

This verse from *The Bad Child's Book of Beasts* by Hilaire Belloc is, indirectly, a relevant comment on nature in the city. Driven to extinction in the seventeenth century, the Dodo, together with numerous other extinct, rare or unusual creatures, were mounted and preserved for posterity in city museums. At one time museums and zoos provided people with the only opportunity of seeing animals they had read about in books. They were collectors items, curiosities to be marvelled at for their size, ferocity, brightly coloured posteriors or strange shapes. Today the animals, mouldering and dusty in their glass cages, are depressing reminders of an growing number of species that have vanished from the scene. Zoos have become more sophisticated in the way animals are exhibited, and are more enjoyable as entertainment; and some have breeding programmes that are aimed at preserving endangered species. But they represent a view of nature that is remote; external to human affairs. On this John Livingston comments that

> the role of Nature in the necessary subsidization of the human interest comes into sharpest focus in the use of animals for entertainment. Animals of all sorts, both wild and domesticated, are pressed into service for this purpose.... In my view, zoos convey and reinforce not only the 'us' and (undifferentiated) 'them' bifurcation of the living world, but also contrive to feed and nourish the fundamentalist myth of absolute human power and control.[2]

Television has raised the educational level of the public with a host of informative programmes on the nature of the living world. But one may question whether the urban experience of nature, and the role of design, is still not largely focused on the exotic bird in a space frame cage and the elephant and tiger secure behind the well-designed moat. Most of our knowledge comes to us second hand through the media. Direct contact with nature and the animal world, apart from resident starlings and pigeons, is confined to non-urban, or for that matter, non-local experience. It is to be had from the weekend at the cottage, or, until relatively recently, on occasional school visits to the rural interpretive centre. Most of us know little about wildlife in the places where we mostly live.

Can the city itself provide us with a direct experience of wild nature? To what extent does the city provide habitat for wildlife? How are these questions important to the quality of the urban places we live in? These are some of the concerns that will be explored in this chapter. To do so, we must review some aspects of natural process with respect to food and habitats, how these are altered by the city, the specific requirements for sustaining wildlife species in various urban environments, and how cities have shaped attitudes and perceptions towards the non-human life that shares habitat with us.

NATURAL PROCESSES

Woodland and forest, grassland and meadow, marshes and water, are the habitat for wildlife. The diversity, structure and continuing evolution of plant communities, their interaction with landform, soils and climate, dictate the diversity and stability of wildlife populations.

The layering or structure of forest vegetation provides distinct environments that support different groups of species. Some feed and breed on the forest floor, some inhabit the understorey and others live in the canopy. Warblers have been found to populate well-defined nesting places according to species, thus enabling a large variety of species to occupy the same forest.[3] Different plant associations provide places for different groups of species. Plant succession produces, over time, a range of habitats from open field to mature forests. Each successive stage is home for different associations of insects, birds and animals. Studies in Maine showed how different species of birds are attracted to these different habitats.[4] Open land was populated by savannah sparrows, song sparrows and bobolinks. In low brush these were replaced by field sparrows, Nashville and chestnut-sided warblers. Pioneer forest attracted ovenbirds, the succeeding evergreen warblers. Woodpeckers and kinglets appeared with the climax forest. Similarly, in aquatic environments, the increased productivity of a lake as it proceeds from an oligotrophic (nutrient-poor), to a eutrophic (nutrient-rich) condition, attracts an

Blackburnian warblers; hemlock tops

Redstarts; sugar maple

Magnolia warblers; lower hemlock branches

Chestnut-sided warblers; low shrubs

Ovenbirds; the cover of the forest floor

a *Different birds live at different levels of the forest depending on their nesting and feeding habitats. Generally the greater the layering of foliage on a vertical profile the greater the diversity of species*

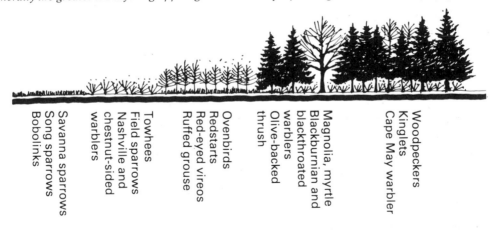

Savanna sparrows
Song sparrows
Bobolinks

Field sparrows
Nashville and
chestnut-sided
warblers

Towhees

Red-eyed vireos
Ruffed grouse

Ovenbirds
Redstarts

Olive-backed
thrush

Magnolia, myrtle
Blackburnian and
blackthroated
warblers

Woodpeckers
Kinglets
Cape May warbler

b *Rojer Tory Peterson's study of bird fauna in Maine showed how different species inhabit different successional stages of a forest from open land, low brush, pioneer forest, mature spruce forest*

Figure 4.1 *Habitat: the layering or structure of forest vegetation provides distinct environments for different groups of species*

Source: *Peter Farb*, The Forest, *New York: Time-Life Books, 1963*

increasing number and variety of wildlife species. Thus wetlands, being highly productive ecosystems, provide habitat for vast numbers of birds and other wildlife. Places that have many different plant associations tend to be richer in species than those that have only a few.

The composition and numbers of wildlife species are also affected by other factors. The edges between one habitat and another are often more diverse than the interior of the habitats themselves. At the same time large-scale interior habitats are essential for some species that would otherwise be vulnerable to edge predators. Continuity of habitat provides essential migratory routes and helps maintain wildlife populations.

The disturbance of the natural environment through human activity sets up imbalances in plant and animal communities. Equilibrium is maintained by an elaborate system of checks and balances. The loss of habitat on which a species depends for food and shelter and to breed may mean that it has to adjust to the new conditions or abandon its chances of survival. The more adaptable species may survive and flourish. Those less adaptable may disappear. Some modified landscapes where woodland, field, scrub, wetland and open water are associated may enhance species diversity. The old complex agricultural patterns of woodland, hedgerows and fields of Britain and Europe have traditionally been rich in wildlife and in historical associations – habitats that have been severely simplified by industrialized farming. The value of hedgerows as a wildlife resource (and for their historical associations with ancient parish boundaries) has become increasingly recognized in Britain, and is reflected in grant-aid schemes for hedgerow planting and management begun in 1992.[5]

URBAN PROCESSES

Urbanization and wildlife

Urbanization has radically altered both natural habitat and wildlife communities. Studies in the United States of the effects of the urbanization of agricultural land on wildlife have documented the changes that take place.[6] Farmland, field and woodland species declined drastically as urbanization advanced and as suitable habitat was reduced. A few species, however, increased dramatically. House sparrows and starlings, virtually absent prior to urbanization, became the most abundant species. The total number of bird species declined, but the total bird population increased. It has been found that there are many more birds in cities, in a numerical sense, than in rural areas outside. The primary effect of land use changes is to fragment forested and other types of habitat, and to convert extensive areas into isolated islands within predominant urban environments. This disrupts the functions of natural areas and inhibits wildlife interaction and gene flow

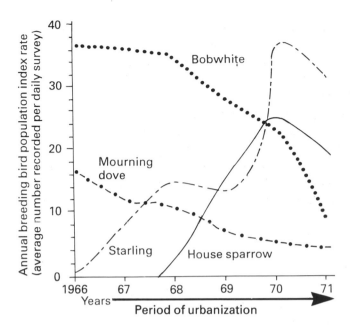

Figure 4.2 *The impact of urbanization (new city of Columbia, Maryland) on bird populations*

The typical farmland species such as bobwhite and mourning dove showed rapid decline. Starlings and house sparrows, virtually absent from the area before development, showed the most striking population increases

Source: *Information for the graph from Aelred K. Geis, 'Effects of Urbanization and Types of Urban Development on Bird Populations', in* Wildlife in an Urbanizing Environment Symposium, Co-operative Extension Service, University of Massachusetts, Amherst, June 1974

among habitats. The concept of island bio-geography has amply demonstrated that large islands support more species of plants and wildlife than small islands. It is now realized that the same principle can apply to habitat patches in terrestrial environments. Forest fragmentation is now considered to be one of the most important environmental issues by many researchers.[7] Taken as a whole, the urban environment is a patchwork of many habitats. The relative numbers, distribution and diversity of animals and birds in various parts of the city are directly related to the diversity, area and structure of vegetation, which determine habitat quality. The plant groups discussed in the previous chapter, therefore, provide a useful guide for an understanding of this question.

The cultivated landscape

The habitat of heavily built-up areas is an environment of buildings, paved surfaces and cultivated spaces. The ornamental and biologically sterile tree and lawn landscapes of downtown parks offer little in the way of food, shelter or breeding places. The herb and shrub layer of natural woodland becomes the paved or grassed floor of the city, trampled by people and permanently hunted by cats and dogs. The intermediate layers have been

Plate 4.1 *Natural habitat structure: complex associations of forest, meadow, wetland and open water*

Plate 4.2 *Changes of habitat structure*

A hostile environment for native wildlife is created when turf is substituted for natural ground layers, intermediate layers are eliminated, when the canopy disappears and when native plant species are replaced by non-food producing ones. Diverse native populations are replaced by a few exotic species

eliminated. The canopy also disappears where trees have been planted as separated individuals. Ornamentals often lack insect life or fruit on which birds can feed. A Swedish study reported that the number of recent and threatened extinctions of higher animals in Sweden is directly attributable to urbanization and the tendency to manicured parks.[8] The disappearance of native animals and birds favours a few aggressive species that are readily adaptable to this habitat. Thus pigeons, sparrows and starlings all thrive in large numbers, feeding off the city's wastes and living and breeding in buildings. All have been introduced. In North America, the house sparrow was introduced in the mid-nineteenth century. The starling was imported and liberated into New York's Central Park in the 1890s. Old world rats and mice arrived in ships accidentally. It is estimated that there may be as many rats in a large city as there are people.[9] The ancestral form of the domestic pigeon is the rock dove. Urban structures, being a human-made substitute for its original windswept Scottish cliffs, provide it with ideal places to roost and nest.[10]

As the density of building reduces beyond the city core, however, more vegetation survives, or is planted along the streets and gardens of residential areas. Robin, redwing blackbird and grackle are common birds in urban gardens and local parks. Even though these are heavily populated by people and domestic pets, the more adaptable native species have gained a foothold. The North American black squirrel, raccoon and skunk have flourished. Raccoons dine on the contents of garbage cans and take up residence in attic and chimney. Ducks and geese in many waterfront parks have become resident and proliferated, feeding in summer on grass and handouts from park visitors, and wintering on patches of open water, created by an industrial plant or generating station. Other species, not normally tolerant of urban conditions, have taken to the city, where the warmer climate and altered habitat provide the right environment for survival. Rooks have formed colonies and bred in urban areas in England; the main shopping area in downtown Cheltenham is an example, with people inhabiting the ground plane, the birds the tree canopy above. Hawks prey in most cities on the small animals that inhabit waste places. The English black redstart has adapted to urban industrial areas where it now chiefly lives.[11]

Native habitat areas

As we saw in the last chapter, many cities still retain elements of native vegetation diversity which can be found in ravines, cemeteries, railways, campuses and waste places. Woodlands of mature and dead trees, open grassland, scrub and young growth, water and wetland, are often present, supporting a variety of native birds and animals. Whereas prairies are dominated by native plant species, old fields contain a large proportion of non-native plants, with introduced species such as clovers, Ox-eye daisy, common St

John's-wort and cinquefoils. In the city centres about half the vascular plants are aliens, arriving with either human or non-human assistance.[12] In the abandoned industrial areas of the Port of Toronto, for instance, open field habitats provide an important food source for at least twenty-seven species of butterfly recorded in the area.[13] A major problem, however, is that these remnant environments are often islands, disconnected from the rural landscape and from each other by the expanding city. And so, where they remain isolated and small, the chances of maintaining high wildlife diversity, particularly animals, are relatively small.

Linkages

Cities are not closed environments, but are connected to rural areas through natural and human-made corridors. Natural corridors include streams and rivers bordered by vegetation or steep banks. Human-created ones include railway connections, canals, highways and transmission lines. These corridors have greatly influenced the migration and perpetuation of wildlife in cities. They maintain the links between natural habitats, parks and the open countryside, and have increased uncommon or non-tolerant species. The ravine system in central Toronto still supports foxes, a relatively non-tolerant animal. London's corridor routes permit movement of animals in and out of the built-up area. Within a 32 kilometre radius of St Paul's Cathedral 314 vertebrate species were recorded by the London Natural History Society between 1960 and 1970.[14]

Problems and conflicts

The imbalances and stresses created by urban activities have many repercussions. Diseases transmitted from wildlife to humans and from humans to wildlife occur more easily when people and animals share limited space in the absence of natural controls. Some diseases are carried by wildlife and transmitted to people. For instance, rabies is widely distributed in rural areas in North America, skunks and foxes being the principle maintenance hosts. The free movement of animals into cities makes the disease a continuing threat to urban and rural people and is expensive to control. Salmonella occurs in English sparrows during cold weather and can be transmitted to pets and people.[15] Some diseases may be carried by pets as well. Toxoplasmosis is particularly infectious to children and is picked up from the faeces of domestic cats in sandboxes and play areas.[16] Urban pets create problems of health and security in crowded city conditions. Feral or stray dogs have been known to attack people, defecation is injurious to plants and is a major source of stormwater pollution. Most major cities regard this as a health hazard, since many older drainage systems discharge stormwater untreated into receiving rivers and lakes in heavy rainstorms. Consequently, overpopulation and uncontrolled breeding, particularly among strays, have become major management problems. The conflicts between people and wildlife range from questions of safety to aesthetics. Birds are hazardous around airports, squirrels damage telephone lines, pigeons make a mess of building cornices and statues in squares and in Venice they are known to contribute to severe deterioration of stone ornamentation and sculpture on historic buildings. Large flocks of urban gulls can temporarily close down lake beaches because of the pollution they create by defecating. This raises questions about what is valuable or a nuisance, and thus of our perceptions of wildlife in cities.

PERCEPTIONS AND VALUES

Our discussion in the last chapter focused on the inherent conflicts of values and priorities that are reflected in the design and management of the urban landscape. These concerns are equally relevant to our discussion of wildlife since we are dealing with inseparable issues. What makes a pet poodle loved and a house mouse persecuted has the same relationship as the garden rose to the dandelion in the lawn. While the environmental movement is changing urban perceptions in the 1990s, the acceptance of nature is still a function of how it conforms to a predetermined set of values and to what extent it is under control. It is tolerated on our own terms, within the limits of convenience, deep-seated acculturation and aesthetic conventions. The conflicts are heightened, however, with wildlife. Animals, birds and insects are more difficult to control. Their presence is more obvious. They are potentially more damaging to human health and welfare. Getting rid of aggressive weeds is less frustrating or demanding than trying to stop the raccoons from spreading the contents of a rubbish bin over the street on pick-up day. Those same raccoons make charming pets as babies, but fully grown they are a menace in the attic or with the pet goldfish. Pigeons being fed in the square have become an inevitable part of the urban scene and a favourite subject for photographers, but in and around buildings they are messy and a nuisance. Life in the city has not been conducive to promoting an understanding of natural cycles. Indeed, it is ironic that wild animals, like most species in nature, have value when they are rare or endangered, but not when they are abundant or successful, a fact that is reflected in government wildlife and habitat protection policies.

What gives rise to these conflicts and human attitudes towards nature is the subject of much philosophical debate in the literature. More suggests that the literary encounters with wildlife by urban children greatly influence anthropomorphic values and popular misconceptions about non-human beings.[17] Worster has shown that responses to nature over the last two centuries have shifted from the Arcadian view of peaceful coexistence to one that expresses the utilitarian view of nature as a resource.[18] Livingston has taken this further, suggesting that our understanding of nature is informed by an ideological insistence that domination is somehow 'natural'. The social theory of the Industrial Revolution, for example, lay behind Charles Darwin's explanation of evolution as a competitive 'struggle for existence', a still prevailing view among biologists that non-human animals compete among themselves for territory and social rank.[19]

Of immediate importance to the thesis of this book is the search for ways in which the city and nature can be brought more closely together to promote alternative values. First, we can act on the notion that environmental literacy in cities involves an understanding of wildlife as an integral part of natural processes and the relationship of life systems to people, and what it can teach us about coexistence. Second, there is the question of diversity and choice. If choice is one of the key factors necessary to a healthy social environment, then a rich and varied wildlife environment is one way of achieving it, since the diversity of human cultures and interests that make up cities should be reflected in a wide variety of urban places. Third, there is the more utilitarian view of wildlife and plants as indicators of urban health. Monitoring and reproduction of shore birds may give us a clue to the condition of the city's aquatic environment. The presence or absence of lichens is an indicator of habitat condition and diversity and therefore a reflection of its suitability for people. How these concerns become part of urban design must be examined in the light of the strategy outlined for plants and the inherent opportunities the city provides.

ALTERNATIVE VALUES: SOME OPPORTUNITIES

The last chapter suggested that the integration of objectives in land-based rural occupations, when conceived as environmentally sustainable practices, and adapted to the city environment, provides an alternative framework for urban design. The same principles are also applicable to wildlife and to plants. While the preservation of rare or endangered species has become a major preoccupation of government policy, these policies tend to ignore the fact that the maintenance and enhancement of representative associations, those that are common or ordinary, are also vital to the biological integrity of urban nature. They may also be achieved more realistically in culturally shaped landscapes that have become a patchwork of remnant habitats. This view is expressed in the science of landscape ecology that explores the integration of ecology and human activity.[20] It describes how a heterogeneous combination of ecosystems, such as woods, meadows, marshes, corridors and human settlement, is structured, functions and changes. It encompasses all environments, from wilderness to urban landscapes, and includes the distribution patterns of landscape elements, i.e., energy flows in soils, the movement patterns of plants, animals and nutrients, and ecological changes over time. Landscape structure includes three major spatial and visible elements of the landscape: patches (wood lots surrounded by agriculture, or open space surrounded by urban development), matrix (homogeneous areas containing distinct patches within it), and corridors (wooded streams, power line rights of way or transportation routes). The interactions of living organisms and habitat within the city environment provides us with the basis for understanding the significance of nature as a whole. And while the protection and preservation of important natural places is where we begin the process of reshaping the city environment, we continue with opportunities for their restoration and creation. These considerations form the basis for the following sections of this chapter.

Natural areas and fortuitous habitat

The habitats with the most obvious wildlife potential are the rivers, streams, canals and pre-urban landscapes that still exist within the city's boundaries (Chapter 3). Those that retain sufficient species and structural diversity of vegetation and are linked to larger habitats are likely to have the greatest faunal diversity. They are, therefore, among the city's richest and most precious places. But the very process of urban growth involves interference with natural successional processes and the creation of a complex patchwork of wildlife habitats. It is these that provide the unexploited opportunities for urban design. They are to be found in the less obvious places that lie behind the city's façades and public travel routes, and that often go unnoticed.

Some are around us in the patches of regenerating land and vacant lots that have been colonized by the plants of the city. Associations of plants and animals are everywhere in evidence. Weedy places abound with voles and mice together with the hawks that feed on them. Dock leaves are food for the painted lady butterfly; stinging nettles are food for the larvae of the red admiral butterfly;[21] monarch butterflies associate with milkweed. Old rooftops may combine standing water and soil and patches of vegetation that become fortuitous wetland habitats for visiting and resident birds. The window ledges of high-rise towers may become a nesting place for peregrine falcons. The natural or contrived impoundments that result from stormwater run-off in many urban areas create new wetlands and wet meadows. Explore the old industrial sites in many a downtown area and you will find unexpected and surprising landscapes behind the chainlink fences, where a combination of a poorly drained site and the natural colonization of aquatic plants have created ideal marshy breeding places for musk rats, ducks and geese.

We must look to those places where energy is concentrated to find other rich wildlife habitats; to environments that have been created by the processes and functions of the city. The waste heat that is dumped into lakes and rivers from storm drains, sewage treatment plants and electricity generating stations, while usually polluted, maintains open water in cold climates and an aquatic habitat for shore birds. Rubbish disposal sites attract rodents and other small mammals and gulls. The availability of food, shelter or heat may provide a breeding place for a species of bird, or benefit migratory flocks and winter visitors. Lagoon systems for sewage treatment adopted by smaller municipalities concentrate nutrient energy and attract large numbers of wading and shore birds, creating habitats that are often richer in species and bird diversity than undisturbed shorelines.[22] As urban habitats, lagoons provide a place for native species of wildlife that naturally inhabit marshes and ponds, and can be seen as one of the city's greatest interpretive and educational resources for naturalists, school groups and the community at large.

A scientific study carried out in 1976 in the regional municipality of Ottawa – Carlton, Ontario, compared the birds inhabiting six of the city's sewage plants with an adjacent stretch of natural shoreline along the Ottawa River. Its purpose was to describe the various sites with particular emphasis on the type of sewage treatment used and how this might affect the diversity and numbers of shore birds attracted to the site. Systematic censuring of each area provided important information on the variety and abundance of shore birds passing through the region during spring and autumn migrations. Table 4.1 shows clearly that birds preferred the lagoons to the natural shoreline. This was so even though available space was more limited in the lagoon system. An important reason for this was the greater abundance of food in the lagoons; another relates to habitat diversity.[23] Thus, far from being 'wastelands' in the city, places that are considered necessary but unmentionable by polite society, the sewage treatment lagoon is a highly

Table 4.1 *Comparison of visits by birds to sewage lagoons and Ottawa Bay (no. of species)*

| 22 July | | 29 July | | 5 August | | 12 August | | 19 August | |
SL	OB	SL	OB	SL	OB	SL	OB	SL	OB
15	6	10	6	18	8	12	11	19	12

| 26 August | | 2 September | | 9 September | | 16 September | |
SL	OB	SL	OB	SL	OB	SL	OB
19	11	18	11	19	12	15	9

Notes: SL: sewage lagoons.
OB: Ottawa Bay – natural shoreline.

Source: Peter J. Hamel, 'Wastewater Treatment and Shore Bird Ecology in the Regional Municipality of Ottawa – Carlton', unpublished student paper, Faculty of Environmental Studies, York University, Toronto, 1976

productive area, lending considerable diversity to the urban environment and thus of great potential value to the study of urban birds and animals. In the face of the widespread disappearance of wetlands in cities, it is also significant that these human-made places have, in some instances, created new habitats, replacing the natural ones urban processes have destroyed.

Wildlife and regenerating waterfronts

The Outer Harbour Headland on the Toronto waterfront is a spit of land nearly 5 kilometres long enclosing the city's outer harbour on Lake Ontario. Begun in 1959, it was intended to accommodate the increased shipping expected from the St Lawrence Seaway. The shipping boom, however, failed to materialize, and for many years the alternative future of the Headland was a source of continuing debate. There were plans to turn it into an aquatic park, complete with hotel, marineland, sailing clubs and parking; visions of constructing a maritime pioneer village, with wooden quays, cobblestone streets and shops.

From the beginning, however, the Headland started to evolve into a complex natural environment and is a fascinating example of the regenerative processes of nature. Its soils consist of an unconsolidated mixture of fill and rubble brought from city construction works and from sand dredged from the lake bottom. This sand was dumped along the sheltered side of the lake, forming four new low-lying embayments and protected lagoons. Over the years on the unprotected lake side, water, wind and wave action have been busy grinding concrete, bricks and stone along the shore into rounded pebbles. Up to 1977, when I recorded its evolution over eighteen years in the first edition of this book, a total of 152 vascular plants had naturally colonized the barren soils, brought by wind, birds and people. Of these, eighty-eight represented introductions into southern Ontario with the remainder being native.[24] Some were rare within the metropolitan Toronto area and two species were identified and recorded by botanists for the first time in the region – golden dock (*Rumex maritimus*) and sticky groundsel (*Senecio viscosus*).[25] By 1992, some 400 species had been identified in seven major plant communities,[26] including two nationally rare, one provincially rare and four regionally rare plants. A forest of cottonwood appeared early in the successional process, and by 1992 had reached a height of 10 to 12 metres.

This evolving landscape soon began to attract migrating, nesting and wintering birds and became a significant staging area and migration corridor. By 1976 some 185 species had been sighted at the Headland. Of the mammals, raccoon, skunk, muskrat, rabbit, Norway rat and groundhog had been recorded.[27] With spectacular rapidity the Outer Harbour Headland became host to great numbers of nesting gulls and terns. In 1973 ten

a *Plan of Outer Harbour Headland, Toronto Waterfront*

b *Outer Harbour Headland vegetation communities*

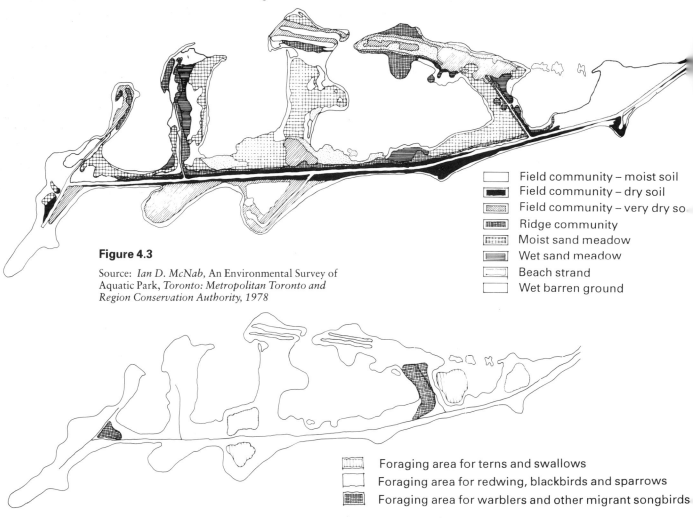

	Field community – moist soil
	Field community – dry soil
	Field community – very dry so[il]
	Ridge community
	Moist sand meadow
	Wet sand meadow
	Beach strand
	Wet barren ground

Figure 4.3

Source: *Ian D. McNab*, An Environmental Survey of Aquatic Park, *Toronto: Metropolitan Toronto and Region Conservation Authority, 1978*

	Foraging area for terns and swallows
	Foraging area for redwing, blackbirds and sparrows
	Foraging area for warblers and other migrant songbirds

Figure 4.4 *Foraging areas for birds, Outer Harbour Headland, 1978*

Source: *Ian D. McNab*, An Environmental Survey of Aquatic Park, *Toronto: Metropolitan Toronto and Region Conservation Authority, 1978*

Plate 4.3 *The Leslie Street Spit from the air, City of Toronto*

An extraordinary emerging productive habitat for fish, animals and birds, protected as an urban wilderness

pairs of ring-billed gulls attempted to breed. By 1977 some 20,000 pairs were nesting on the peninsula, growing in 1982 to 80,000 pairs, along with other breeding birds that included the herring gull, common tern and the rare Caspian tern. And by 1992 the total number of bird species had grown to 290, of which forty were breeding at the site. Five species of waterbirds nest in significant numbers: the herring gull, common tern, black-crowned night heron, double-crested cormorant and ring-billed gull – one of the largest colonies in the world. The Caspian tern, for reasons unknown, has disappeared. But in April 1993 coyotes were sighted on the Headland, travelling down the Don River valley corridor to take up residence. Conservation Authority staff found evidence of den sites, which suggested that a pack was established and breeding.[28]

This remarkable place has become one of the most significant wildlife habitats in the Great Lakes region in an environment where industrial growth has destroyed many of the habitats that birds require, and has rendered others toxic. Before large-scale harbour developments began in 1912, Toronto's waterfront was one of the richest and largest marshes on Lake Ontario. These were filled in to create the harbour, industrial lands, railways and expressways that are the legacy of the city today. But the creation of the Headland has, by accident, recreated some of the very habitats that were destroyed earlier in the century. Its existence has permitted much wildlife to recolonize. Without the construction of this peninsula biologists believe that many species such as the common tern would no longer be nesting in the Toronto area.[29] And it is almost certain that the coyote, a prime carnivore, would not be residing in downtown Toronto.

The opportunities that occur fortuitously as a result of urban processes have great relevance in our search for an alternative urban design philosophy. I have already suggested that the principle of diversity offers variety of place and social opportunity. It is apparent, if one begins to observe the city with an ecological perspective, that this is occurring unwittingly in many places and in many ways. It needs only to be recognized and used to advantage. I have also suggested in Chapter 1 that the principle of least effort is a relevant and important objective. It is perhaps best summed up in an early leaflet distributed by the 'Friends of the Spit', the group of citizens who, since 1976, have led the campaign to have the Headland protected as a nature reserve:[30]

- What do we want . . . and how much will it cost?
- Keep the place as it is without any major development.
- Let it develop naturally into an increasingly secure wilderness.
- Jettison the (Metropolitan Toronto and Region Conservation Authority) master plan approach.
- The nice thing is that to do these things will cost very little. And millions of taxpayer dollars can be spent instead on more pressing social and economic needs.

Further developments since the 1980s

In 1987, in response to public pressures to protect this urban wilderness the Metropolitan Toronto and Region Conservation Authority (the agency responsible for managing the Headland) approved a master plan for the area. In addition to protecting its key habitats and wildlife, it also proposed the establishment of sailing clubs and a major interpretive centre together with parking for an overall estimated cost of some Canadian $5 million. Pressure to revise this plan was exerted by the Friends of the Spit, together with similar recommendations from the Royal Commission on the Future of the Toronto Waterfront and the City of Toronto Planning Department. Consequently, two years later a revised plan was brought forward that largely met public insistence on a car-free, no development, wilderness environment that should be left to evolve on its own. Proposed management strategies included, among others, the following recommendations from the Friends of the Spit:[31]

- The best philosophy for the Spit is to 'Let it Be'. The least intervention scheme will be the best.
- An ongoing review committee be established to review and monitor habitat and species management proposals.
- Habitat and species management shall only be undertaken on the soundest of academic experimental conditions (an example are the barges floated out during the breeding season for nesting common terns that has assured their continued success).
- If plantings are to occur, then native species should be used.
- The Spit is always an evolving laboratory. It has evolved to its present vegetative state by natural seed and root transport. Therefore, speeding up natural processes is not desirable.
- Woodland clearings 'for aesthetic reasons' is inconsistent with a wilderness park concept.

Some strategies have been put in place to ensure protection of the birds while encouraging walking, cycling and ornithology. The main route that runs along its entire length confines most people naturally to a specific corridor. The main habitat and breeding areas that can be reached by frequent informal pathways are posted during the breeding season with 'no entry, sensitive site' signs that explain the nature of these habitats, and ask people to go no further during the nesting season. Small open barges are floated out into the lake every spring for breeding terns, allowing people to observe from the land, and also preventing the terns from being crowded out by gulls. The management proposals put forward by the Friends of the Spit raise interesting issues. First, there are automatic assumptions by controlling agencies that significant interventions and associated costs through master plans are an essential mandate for planning urban lands. In contrast, citizen groups composed of committed, well informed and scientifically credible people are increasingly insisting that natural processes should, on occasion, be left to evolve on their own with a minimum of interference. Second, it is also assumed that *aesthetic* intervention is a requirement of such plans; an aesthetic based primarily on dictatorial and preconceived agendas learned in design school. As the Friends succinctly have observed, 'Manipulating vegetation "for aesthetic reasons" (*and whose aesthetics?*) will negate the uniqueness of the experience, and require unnecessary expenditures.'[32] Third, that effective, knowledgeable action on the part of citizen groups (such as the Friends of the Spit), is becoming increasingly effective in sponsoring and protecting natural areas in cities. As a consequence of these public actions the Spit has continued to thrive and provides an unparalleled wilderness experience for the 50,000 people who visit the site every year. Fourth, its experiential and scientific value to the city

Plate 4.4 *A place to contemplate and experience natural forces in an urban setting*

Plate 4.5 *A colony of ring-billed gulls*

The gull habitat is being taken over by woodland, but more is being created by fill extensions to the Headland

lies in the opportunity to see a complex natural community develop from the ground up. Its protection has been achieved both by public pressure and by default. As David Crombie, the Commissioner for the Royal Commission on the Future of the Toronto Waterfront, aptly put it 'While all the debate was going on, nature decided the issue.'[33]

Amphibians and reptile habitats

Another aspect of the wildlife scene in urban areas, and one that is often ignored, is the habitat that the city does and could provide for frogs, snakes, lizards, turtles and other amphibian and reptile species. These animals, like the more obvious ones that we have been examining, not only rely on remnant areas of undisturbed habitat, but also on altered sites. Riverine marshes, for instance, may support many species of turtle; the stone filled wire baskets forming urban stream banks may be habitat for garter snakes; old gravel pits and fortuitous water impoundments are often places where frogs, toads, turtles and snakes can be found. While amphibians and reptiles have been among the least recognized of the urban fauna, it will be self-evident that their continued existence in viable populations is critical to the maintenance of species diversity and stability. In a study of chorus frogs in the United States it was found that their life expectancy was about four months with 95 per cent of the population failing to survive the winter.[34] Bob Johnson observes that 'this relatively high rate of mortality may be typical of most of our amphibians which are nurtured in the very productive waters of marshes and swamps. As the frogs are eaten, the energy they have carried from the swamps is passed on to other organisms that otherwise could not exploit the richness of these areas'.[35] Thus to save wetlands means more than saving the amphibians and reptiles that one finds there. It also means that one can save all the trapped energy for use by a whole chain of plants, invertebrates, birds and mammals elsewhere.

PLANNING AND MANAGEMENT ISSUES

A strategy for urban wildlife habitat

The naturally occurring wildlife places that we have been examining provide the inspiration for a purposeful design strategy. This takes three essential forms: identifying and planning for what is there now (existing habitats within built-up urban areas), restoring already degraded habitats, and identifying what could be there in the future (a regional planning approach for future urban growth that identifies important habitats that should be protected and integrated into open space systems).

Identifying and protecting what is there now

The city has innate opportunities for complex wildlife habitats. As agriculture has industrialized, there has been a tendency for rural areas to become ecologically less complex, and the city more so. This provides us with the basis of a design strategy. Its fundamental objective should be the enrichment of the city's existing natural wealth; to capitalize on what is inherently there. This is, in fact, one of the significant attributes of the cities of the western Netherlands. The great diversity and richness of its park system, waterways and streets, the mixture of old and new, the contrast between naturalized areas, productive and manicured ones, is in marked contrast to its agricultural hinterland. This is a modern, food-producing landscape of big machines and unlimited views, that has little ecological or visual richness, or recreational interest. The Dutch have embarked on a programme of enrichment of the urban landscape which counterbalances this contrast between urban and rural environments. While limited space is the crux of the problem for the Dutch, elements of such a strategy are highly relevant to other cities.

At a broad level of planning, the urban environments that have special value for wildlife must be integrated into the spatial networks of the city. This must include not only the recreational parks, but essential natural and human-made connections and the exterior environment as a whole. This is particularly significant since many of the most ecologically important urban places are not often included in park planning. The identification of habitat types provides a start for locating highly valued places requiring restricted or carefully controlled access. However, the protection of Environmentally Sensitive Areas has often been based on drawing an artificial zoning line around them, while ignoring adjacent natural areas that lack 'special' features. Consequently, many of the protected 'special' areas are no longer able to support the species and features that gave rise to the designation in the first place.[36] Other habitats may be identified in descending order of significance or sensitivity to disturbance (see pp. 111–12). Special areas include:

- Remnant rural landscapes enclosed within the city's boundaries that maintain an existing community of wildlife species or harbour locally rare or unusual species. Examples include woodland associations, old field and meadow, watercourses, marshes, natural corridors.
- Human-influenced places that have high potential. Examples are the city's sewage treatment lagoons, abandoned naturally regenerated lands, industrial lands, aquatic wintering sites.
- Areas of potentially high wildlife significance and sensitivity where the management of stormwater and wildlife habitat can be integrated. These might occur in

Figure 4.5 *Potential wildlife habitat types*

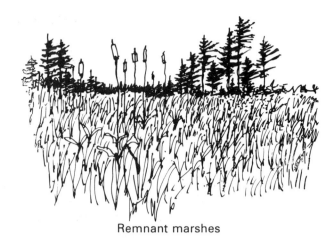

Remnant marshes

Remnant woodlands and corridors within the city's boundaries

a *Remnant rural landscapes*

Aquatic wintering grounds

Abandoned industrial sites

b *Human-made resources*

Major city parks that have wildlife potential

c *Wildlife habitat integrated with stormwater impoundments and watercourses*

Water courses

Golf courses

d *Places where wildlife habitat and human activity may be integrated*

City parks and institutional lands

e *Places not barred to the public, but having little or no direct use*

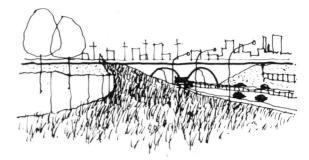

Road edges

Disused canals

Expressway interchanges

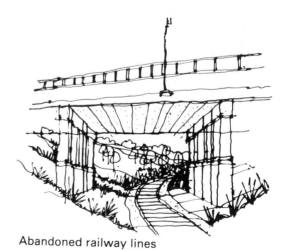

Abandoned railway lines

association with residential open spaces, major city parks, wetland reserves, floodplains and watercourses.

- Other areas less sensitive to a human presence. These may include open spaces with a wildlife potential that could, with management, accommodate a combination of human activities and wildlife. They might include public parks, cemeteries, golf courses, institutional lands, industry, public works property and related areas.
- Linear connections linking habitat areas. These may include river and stream corridors, escarpments, continuous woodland, transmission corridors and pipeline rights of way.

The ability of wildlife to survive urban pressures depends on the complexity, productivity, size and shape of habitat. It also depends on the intensity and kind of public use and the varying degrees of restriction imposed on the site. The best habitats are often those with the greatest impediments to human use. For instance:

- Places that are open to the public but have little or no direct access. These include travel corridors, such as main roads and urban expressways, abandoned railway lines and canals.
- Places that are restricted to the public for social or security reasons. These include properties and gardens, public works property such as sewage treatment plants, water reservoirs and some high-security industries such as refineries and airports, electricity generating plants and many flat rooftops. Incorporating these into a city-wide wildlife network is critical to the overall enrichment of the city.

Restoring what has been impaired or destroyed

At the level of design and management, the task is threefold: first, to enhance potentially significant wildlife habitats; second, to create new habitats where urban land offers potential opportunities for this to occur; third, to restore appropriate connections between habitats. Several factors are relevant here and involve a number of measures:

- Generally the larger a habitat is the better it is for wildlife. At very small sizes, the increase in species with increasing area is almost exponential, but the rate of increase begins to taper off once the habitat is larger than 5 to 10 hectares. Above this size, the number of plant species may increase gradually with increased size, but additional species of birds and mammals may not appear until certain size thresholds are reached.[37]
- Even the smallest places provide habitats for some wildlife species. For instance, areas as small as 0.1 hectare will attract common breeding birds depending on the

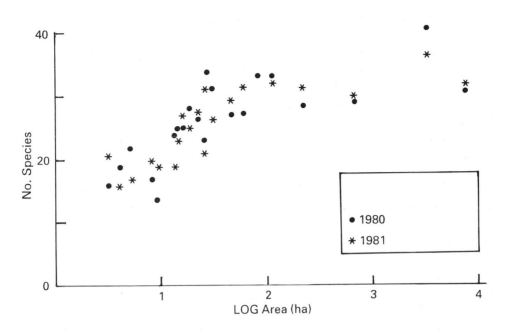

Figure 4.6 *Species-area patterns in forest birds near Ottawa*

According to a study by Freemark the number of bird species increased significantly with forest area. Forest size explained half or more of the variation among forests in the number of bird species, bird pairs and pairs per species. Components of habitat heterogeneity were of secondary importance

Source: *K. Freemark, 'Landscape Ecology of Forest Birds in the North-East', in R.M. De Graaf and W.M. Healy, 'Is Forest Fragmentation a Management Issue in the Northeast?', Society of American Foresters Annual Conference, Rochester, 1986*

type of habitat. All are species that have adapted to potentially high rates of disturbance, predation and parasitism. An area of 4 hectares of woodland is perceived as pure edge with no interior habitat, but 100 hectares of woodland is large enough to provide habitat for almost all forest-dependent birds except those with very large home ranges.

- Planting should provide a diverse vegetation structure for breeding, cover and food sources. A key factor in habitat creation is the need for large-scale patches (woodland, wetland, etc.) for interior species. For example, ornithologists estimate that the area up to 100 metres from a forest edge is dominated by edge species. Interior species start to appear 100 to 180 metres from the edge, but deep-woods species are seldom seen closer than 300 metres from the edge.[38]
- Maintaining vegetation diversity where it exists. For instance, existing old and dead trees, ground litter, brush piles and thickets in woodland parks and open spaces

provide appropriate conditions for a variety of species. The frequent practice of clearing and tidying up, where safety is not involved, minimizes natural diversity and needlessly increases maintenance costs.

- Providing nesting boxes, bird feeders, water sources.
- Providing places and paths for people to avoid conflicts and disturbance.

Habitat types

PRIVATE PROPERTY

The long-term stability of private property is a question that has considerable bearing on wildlife and one that makes urban situations different from rural ones. Some types of property tend to imply a vested interest in the protection of natural areas within the context of human use. Old cemeteries grounds, where benign neglect has permitted them to evolve into a complex pattern of vegetation, are potential habitat for a variety of wildlife. Golf courses, often located in valley lands, scenic landscapes or wetlands, are considerably more problematic. Wooded slopes, copses, water, golf greens and brush do create a complex habitat and they incorporate large areas with a low intensity of use. However, there are environmental problems. Fertilizer and pesticide run-off from turf grass and seepage into streams and groundwater, while reportedly negligible during the summer growing season, occurs in autumn and winter.[39] A US Environmental Protection Agency study in the mid-1980s found eight of sixteen pesticides tested in the groundwater beneath four Cape Cod golf courses.[40] Golf courses can often disrupt streams and vegetation, and fences inhibit both human and animal movement along valley bottoms. They are, however, among the fastest growing recreational activities in urbanizing areas worldwide. A report by the Royal and Ancient Golf Club of St Andrew's (the game's controlling body in Britain) has concluded that some 690 extra courses will be needed in the UK by the year 2000 to satisfy demand beyond the 1,800 currently in existence.[41] Given a universal demand, positive steps towards creating an environmentally benign facility can address the physical and long-term ecological concerns. These might include such actions as creating more naturalized chemical-free landscapes, encouraging diverse plant associations, protecting streams and fish and establishing linkages.

Private residential property that incorporates or creates natural habitat, has great potential for attracting wildlife. As Chris Baines has shown eloquently, even small gardens can be made to have variety of habitats:[42] with a patch of woodland and understorey to attract warblers, squirrels and birds and mammals that prefer habitats close to the ground, long grass and meadow for butterflies, moths and other insects, a pond that supports bulrushes, sedge and water lilies that attract frogs, toads and dragonflies. Food sources may be provided by planting fruit and berry-bearing shrubs

Plate 4.6 *An invitation for bird watchers in a local cemetery, Toronto*

Considering the highly mown state of the cemetery environment, the place is remarkably diverse, assisted by its connections with a local ravine

and trees, patches of wild or cultivated flowers that are a source of food for seed-eaters in the autumn and winter. He points out that when the individual garden is seen in context with the surrounding landscape of other neighbourhood gardens, with remnant woods and scrub, naturalized railway rights of way and canals, a connecting framework of natural areas becomes possible, linking up to form a wildlife network of travel corridors. The grass-roots Urban Wildlife Group in England has carried out detailed surveys of much open space in the countryside, and has helped local authorities to prepare policy documents to identify places that need protecting. As Baines has commented 'we have helped make nature conservation "respectable"', a particularly important factor in a country where 97 per cent of meadows have been destroyed over a period of thirty years, and where 10 per cent of its Sites of Special Scientific Interest are being degraded every year.[43]

SEWAGE TREATMENT LAGOONS

The potential of sewage treatment lagoons could be enhanced by simple management practices. Nesting boxes encourage birds to breed. Planting low shrubs below lagoon dykes provides cover for land birds.[44] Well sited look-out points, blinds and trails permit observation of birds without disturbance, and control access and movement. Interpretive signs and pictures of birds can, in some instances, enhance visitor experience. This is a frequent practice for wildlife reserves in urban parks in Britain and Europe. Other simple interpretive techniques might include walking trails illustrated by pamphlets or recorded tapes. This latter method is both effective and inexpensive and has been used for a number of years at the outdoor trail system in the Petawawa National Forestry Institute near Ottawa.[45] Distribution and return of materials is controlled and incorporated into the Institute's existing security facilities.

CITY PARK WILDLIFE

Setting aside wildlife reserves can enhance many city parks and reduce maintenance costs as well. The lake in Stanley Park in Vancouver is a favourite place for visiting and breeding birds. Disturbance to breeding swans from people and domestic pets is minimized by incorporating simple fences between the footpaths and nesting sites along the lake edge. The enjoyment and educational value of seeing nesting birds is thus maintained easily and effectively. One of the most famous parks in London, Regent's Park, incorporates a protected wildlife area within its boundaries. A densely wooded island in the lake provides undisturbed sanctuary for many species of ducks and other birds. The island is also home for a permanent colony of herons that live and nest in its mature trees. These normally shy birds have become accustomed to living in a park environment, surrounded by the noise and traffic of the city and the everyday coming and going of park users. The whole of the area surrounding this part of the lake is separated from footpaths by water and a fence. But the sanctuary is close enough to allow people to observe the activities of its inhabitants. To see these great birds flying low over the water, or feeding their young at treetop level in April, with the sounds of the city in the background, is an extraordinary and quite unforgettable experience.

In the urban parks of the Dutch cities, the well-known image of tulip beds and beautifully tended lawns is only one part of a great diversity of habitats that are considered essential to a good park environment. These habitats serve many purposes, including a climatic function, active and passive recreation, education in natural sciences, horticulture, animal husbandry and allotment gardening. In the Hague about 300 different bird species live in or visit the city, with some ninety-five species breeding there. There are five or six local nature reserves of different kinds, including woodland and dune landscapes. One of these, accessible only to members of the Association for Bird Protection, is an area of old dune landscape where green woodpeckers, owls, finches and jays may be found together with wild plants that have ceased to exist elsewhere.[46] It must also be remembered that there has been painstaking reconstruction of natural areas in some cities in the western Netherlands, whereas many other cities have existing natural and human-made features that can be incorporated into the fabric of an open space system. An example is the natural deep valley that runs through the centre of the city of Luxembourg, where in the evening one can overlook the valley and listen to the nightingale singing in the middle of the city.

ROOFTOPS

The flat rooftops of many commercial and industrial buildings are an example of often highly visible, but publicly inaccessible, places. They could, with some adjustment to roof design, provide the ideal sites for upper-level urban wetlands and other habitat types. A few centimetres of water that can pond in some areas and allow marsh vegetation to establish itself could, when designed with the same care as decorative rooftops, provide stop-over places for migratory birds and nesting sites for resident species. The rooftop wetland could, to some extent, replace some of the natural habitat lost to urbanization. There are other benefits that make the urban rooftop a valid expression of an alternative design form. Retention and the controlled release of rainwater to drainage systems reduces the impact of sudden storms on watercourses, a problem that is of increasing concern to many urban municipalities (Chapter 2, p. 39). The presence of water and vegetation at rooftop level assists in the control of heat build-up during the summer months, through direct evaporation into the air and evapotranspiration by plants (Chapter 6, p. 245). The potential benefits to the life of the roof by being protected from continuing expansion and contraction should also not be discounted. Wetland habitat becomes the basis for a vernacular urban landscape and an aesthetic derived from the integration of natural process, technology and design. Roof gardens as natural habitat are common in Europe, particularly in such cities as Zurich, Hanover and Dusseldorf.

INDUSTRY

The principle of visual access but physical separation for wildlife areas is highly relevant to high-security industries. An example of this is the Petro-Canada Lakeshore oil refinery in Mississauga, Ontario. This 225-hectare industrial complex is abutted by urban development on three sides and Lake Ontario on the fourth. In 1975, the refinery underwent an upgrading of its plant. Faced with the realities of having to live in harmony with the communities that surround it, it initiated a landscape plan around its boundaries as a part of a plant modernization programme. The plan, conducted in co-operation with neighbouring communities, included the creation of extensive screening and wildlife habitat with earth berming and planting within the refinery's fenced boundaries. This came about at the request of the community, many members of which were keenly

Plate 4.7 *Lakeshore Refinery (Petro Canada) in relation to the surrounding communities and the Rattray Marsh (bottom left)*

The marsh, creek corridor and refinery link has encouraged animals such as deer into these newly created refinery lands

(Photo: Steve Frost)

Plate 4.8 *The refinery from Meadowwood Park*

The corridor also serves as a visual screen for the houses facing the refinery. Note the mowing line within the park that extends the naturalizing area beyond the refinery fence

(Photo: Steve Frost)

interested in wildlife. People had over the years witnessed the continuing deterioration of the environment where they lived.

Analysis of the larger area revealed some interesting features. The site was close to a lakeshore marsh (the Rattray Marsh Conservation Area) which was fed by a wooded creek that extended to its easterly boundary. Thus a reforestation of this boundary strip would create a natural circular link connecting the marsh with the refinery. The site itself had been disturbed by landfill and other refinery operations. It was, therefore, in poor physical shape. But a natural drainage channel running through the site had created a small area where aquatic vegetation could flourish. There also remained the vestiges of old field communities of wild grasses and flowers from the previously existing agricultural landscape.

The plan that evolved was intended to maximize diversity and create a number of distinct wildlife environments. A pond was dug adjoining the existing wetland, combining open water with an extended aquatic vegetation. The old field community was retained and extended. A variety of small deciduous and coniferous woodland habitats were planted, including the re-establishment of a number of locally native species that had all but disappeared from the area. A linear wooded strip of about 50 metres was planted along the entire eastern refinery boundary to link the wildlife area to the adjacent lakeshore wetland in a continuous corridor.

A critical element was management. One of the local community's goals was to restore habitat and allow it to evolve undisturbed. To achieve this required three long-term strategies: a minimum of interference from horticultural maintenance; little or no human intervention; and continuing regular public meetings.

Minimal interference

An initial period of grass-cutting during the first several summers after installation was undertaken to reduce competition to shrubs and trees. After that all mowing, with the exception of some boundary areas adjacent to residences, was abandoned.

Human intervention

The management plan was based on permitting visual rather than direct access to the natural areas within the refinery boundaries. Boundary fencing therefore remained as security for the refinery operations with no public access to the reserve, except under special supervised conditions. Outside the refinery's security fence and within the existing park separating the community and refinery, a large hill was built to permit wildlife viewing at a distance while excluding the public from entering the area itself. Other views of the pond were designed to permit closer observation by local residents.

Continuing public meetings

These have been held on a yearly, or more frequent basis since the completion of the project between refinery management and the various neighbourhoods to continue dialogue and resolve problems as they arise.

Developments since the mid-1980s

Over the period of its implementation, the various plantings have become established and begun to develop distinct woodland, meadow, open water and wetland habitats. A diversity of wildlife has returned to colonize the area. Muskrat have appeared along with raccoons, foxes, ducks, geese and countless songbirds, and in 1985 deer set up residence within the refinery boundaries, following the natural corridor from the Rattray Marsh and leaping over the 2.5 metre fence to take advantage of the wooded habitat and food sources. The hill in the existing park provides views to the plant associations of the reserve, and to the lake – views not previously experienced. Children use its slopes for tobogganing in winter. The Parks Department have established a naturalization programme between the refinery corridor and the park, thus extending the natural corridor into the park environment. Maintenance costs for the first four years of establishment amounted to 298 person hours per hectare per year. Following this period maintenance costs were reduced to 38, a reduction of almost 800 per cent (see Table 4.2).[47] An initial wildlife habitat management plan was implemented some ten years after its initial creation, involving thinning of some woodland trees to reduce competition and

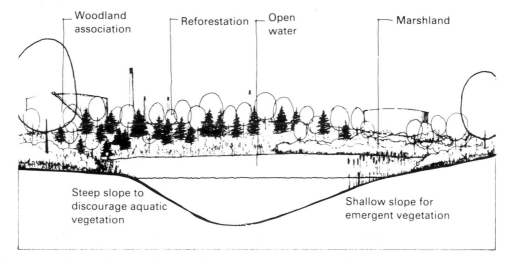

Figure 4.7 *Section through the pond showing construction principles*

Plate 4.9 *The wildlife pond and wetlands and littoral edges beyond, creating habitat diversity and functioning as a source of water in case of fire*

(Photo: Steve Frost)

promote a canopy and ground flora, some additional plantings and meadow seeding and the redefinition of mowing along edges adjacent to residential areas.

This is a place where a practical attempt to recreate a natural habitat in urban surroundings has been made. The association of wildlife and industry is a symbiotic one. Industry has benefited from a landscape buffer between it and the community that needs little or no upkeep. The community has benefited from the educational and enjoyable environment that is available on its doorstep. Both the industry and the surrounding residential neighbourhoods have benefited from the continuing dialogue that began in the 1970s and has continued regularly every since, greatly reducing the sense of confrontation and complaints that once existed.[48] And wildlife has flourished, protected from direct intrusion on its territory by security fences.

Table 4.2 *Summary comparison, conventional and low maintenance landscapes*

Type	Location	Comparison area	Summary maintenance programme	Changes to lower maintenance	Cost (hours/ha/year)
Immature low maintenance landscape					
Reforestation berms (private)	Gulf Canada Clarkson Refinery Mississauga, Ont. (Orr Road and Meadow Wood Road berms only)	8.46 ha (contract area only, same as below)	– medium level of maintenance – grass cutting 3 times a year – trimming, fertilizing, litter pickup, rodent control, pruning irrigation, cultivating	Before manipulation – constant annual programme during first 4 years of establishment (see below)	298 hours/ ha/year[1] *Note*: cost slightly higher than for public open space due to complex trimming and cutting operations
Mature low maintenance landscape					
Reforestation berms (private)	Gulf Canada Clarkson Refinery Mississauga, Ont. (Orr Road and Meadow Wood Road berms only)	8.46 ha (contract area only, same as above)	– low level of maintenance – grass cutting 4 times a year – litter pickup, irrigation, rodent control	After manipulation – cutting on slopes eliminated – no pruning – smaller, less costly machinery	38 hours/ ha/year[2]

Notes: 1 Cost slightly higher than for public open space due to complex trimming and cutting operations
2 Cost substantially lower than for public open space due to reduced cutting. New complex mowing regimes and reduced cutting have created a rich diversity of environments

Source: Glenn A. O'Connor, 'Establishing and Maintaining Low Maintenance Landscapes for Southern Ontario'. Unpublished research report prepared for the University of Toronto, Faculty of Landscape Architecture, 1984

Some other aspects of management

The imbalance of ecological conditions caused by urbanization creates problems as well as opportunities. An example is the vast increase in exotic species of birds and rodents. Bringing technology to bear on environmental problems is often difficult and expensive. In some cases it may be a question of minimizing hazards, or being aware of the problem and living with it: 'We don't destroy wild birds because they carry encephalitis viruses. Instead we try to control the mosquito vectors – by means that will not injure the birds.'[49] There are many situations where a greater balance could be brought to the city environment by a greater respect for natural processes in design and management. There are several examples that illustrate this.

The overwhelming reliance on a few, mostly exotic, species of trees for urban parks and gardens is based largely, as we saw in the last chapter, on horticultural, engineering and decorative criteria. Their relationship to such factors as insect and animal diversity is rarely considered. So, while they will not clog the drains, or cause a problem with the telephone wires, the habitat created is generally sterile. A study of the common trees in Britain, together with their history and the insects associated with them, is revealing.[50] A few are summarized in Table 4.3 and show that the native trees have the largest numbers of associated insects (oak with 284) and those introduced by far the least (plane with zero). From the perspective of wildlife diversity alone, the incidence of insect communities related to plants has some considerable significance for the selection and creation of viable habitat.

In central urban areas, buildings have become bird habitat, but the communal roosting and nesting habits of pigeons and starlings create problems. The large quantities of droppings they produce deface buildings and floor surfaces and are expensive to clean up. For many historic cities like Venice, this is a serious matter. The sculptural surfaces of historic buildings and statues make fouling more obvious. At the same time, one of the great attractions of Venice is its famous pigeons in the Piazza San Marco and local

Table 4.3 *Common trees in Britain*

Tree species	History in Britain	Total no. of associated insect spp.
Oak (*Quercus robur* L. and *Q. petraea* (Matt.) Liebl.)	Native	284
Birch (*Betula* spp.)	"	229
Willow (*Salix* spp.)	"	266
Hawthorn (*Crataegus* spp.)	"	149
Lime (*Tilia* spp.)	Native and introduced	31
Horse-chestnut (*Aesculus hippocastanum*)	Introduced *c.* 1600	4
Acacia (*Robinia pseudo acacia* L.)	Introduced *c.* 1501	1
Plane tree (*Platanus orientalis* L.)	Introduced *c.* 1520	0

Source: T.R.E. Southwood, 'The Number of Species of Insects Associated with Various Trees', *Journal of Animal Ecology*, vol. 30, 1961

commercial photographers, with trained cats at the ready to send the pigeons flying, take full advantage of them. It is ironic, however, that studies have shown that guano protects limestone from sulphur dioxide, the industrial pollutant to which Venice and other ancient cities are highly prone.[51] In modern industrial cities the solution is to design buildings to inhibit nesting and perching. Crevices, cornices, ventilation shafts and similar holes and projections could be simply protected and bird numbers controlled by reducing available habitat.

The use of predators may control or reduce the number of common birds that have either become a nuisance or are a hazard in certain situations. For instance, snowy owls are regularly used at some airports where the danger from birds to aircraft taking off and landing is potentially an extreme safety hazard. Peregrine falcons have been introduced into some cities to prey on pigeons, starlings and gulls.

Mammals also do well in the habitat and plentiful food supply that the city provides. For instance, male raccoons in the wild claim a territory of up to 5,000 hectares where females are allowed to roam with their young. In the city, however, their numbers more than double. A study by the Ontario Ministry of Natural Resources showed that Metro

Plate 4.10 *Raccoons have perfect habitat conditions in many cities: plentiful food from household wastes and habitat in natural corridors*

Toronto has some 10,000 raccoons, roughly sixteen to eighteen per hectare.[52] Toronto has the distinction of being known as the 'Raccoon Capital of North America', but they are found in almost every region of North, Central and South America. In addition, as a consequence of the introduction of a few dozen animals to Europe from North America early in the twentieth century, Germany, France, the Netherlands and parts of Russia now have an exploding raccoon population estimated at 100,000.[53]

Urban contaminants

Less is known, however, about the impact of the city environment on the health of wildlife. The uptake of toxic chemicals and other contaminants through the food chain, from city sewers, the soils of abandoned industrial areas, groundwater, open water, sediments and air, has been detected in the Port Industrial Area of Toronto's waterfront. Studies were conducted on benthic invertebrates across the Toronto waterfront in 1991, and included caged clams, snapping turtles, sportfish and fish eating birds.[54] Overall they showed that all trophic levels in the aquatic food chain are contaminated with toxic chemicals, an indication of relative ecosystem health that has repercussions on human health. Air pollution is also a well-known health hazard to human beings and domestic animals, but its impact on wildlife has been less well researched. Where population measurements have been taken, significant reductions in vertebrate wildlife have been correlated with industrial air pollution.[55] Population censuses of the house-martin in the former Czechoslovakia, for instance, have shown the species to be rare or absent in areas with heavy fluoride, sulphur dioxide, fly ash, cement and nitrogen oxide pollution.[56] In London since the turn of the century, there has been a decline and subsequent return of bird populations to the inner city. This has been attributed to the reduction of the once high levels of smoke and other pollutants in central London.[57] The importance of industrial pollution as a factor contributing to the decline of wildlife, therefore, should not be underestimated and it illustrates the essential links between animal and human health in cities.

In summary, the value of wildlife in the city has to do with the lessons we can learn about the balance of nature and how human life is part of that balance. Study of human and natural urban systems is perhaps the best way of coming to grips with the wider issues of vanishing species and pollution. Hilaire Belloc's poem about the Dodo is, therefore, an apt reflection on our problem and the need for city nature. There remains a perceptual dissociation between urban and rural environments that has radically influenced the way urban people view wildlife and nature in general. Cottaging, camping, hunting and most other leisure-oriented activities that are pursued in the countryside are basically exploitive occupations when there is no long-term investment in nature or the land. Yet our examination of wildlife and its dependence on the other elements of nature's processes shows that many of the city's most biologically productive environ-ments are to found where energy and nutrients are concentrated. And this brings us to the question of how this phenomenon may become a way of regaining an investment in the land. The productivity of land, growing food and our ultimate dependence on the soil for survival strike at the heart of this issue. What implications does food growing have for cities in the light of ecological and conserver values? What are the connections between soil, productivity, energy and nutrients, and the alternative framework for urban design that we have been exploring? These are questions that now deserve our attention and will be addressed in the next chapter.

5

CITY FARMING

•

INTRODUCTION

A 1980 issue of the British *Farmer's Weekly* journal offered some alternative guidelines to the Countryside Commission's Country Code on how urban visitors could cement friendships with country folk.[1] Here are a few of them:

> **Gates:** A modern technique called zero grazing has made all gates redundant, so there is no need to bother with them. Never open a gate if you can climb over it first. If it is locked, take it off its hinges.

> **Join in the fun:** eartagging is a common practice in almost all livestock enterprises. Try catching some animals to see whether they have any tags. A farmer will roar with friendly laughter if he finds they have been removed.

> **Share a good laugh:** no one likes a practical joke better than a farmer. When passing a farm, call in and tell him his sheep have got out. Remember, if they have, never direct him to the right field.

> **Bulls:** contrary to popular belief only one breed of bull is dangerous – the savage Hereford. The majority are warm-hearted amiable creatures loving nothing better than to munch grass and attend shows. When out walking, why not try scratching a Friesian bull behind the ears with a blunt stick? The results can be interesting and rewarding.

While tongue-in-cheek, these comments are a commentary on the basic conflicts between urban and rural values and social perceptions towards the land. To most urban people, the countryside is a recreational resource – a place to escape to from the city. The connection between food and the land on which it is produced has become increasingly remote as an issue that directly concerns urban welfare. The food that appears in the supermarket has little direct connection any more with the fields adjacent to the city. It is instead, dependent on worldwide marketing and distribution networks operating on fossil fuels and based on international trade agreements. But current patterns of consumption and environmental priorities are shifting. One of the indicators of change can be found in the concern for diet, the growth of health food stores, farmers' markets and allotment gardening, that suggest there are signs of a return to home-grown versus 'factory-made' food. In this chapter we are concerned with how the links between people, food growing, 'waste' and city space, can be creatively re-established as a central aspect of environmental and social values.

AGRICULTURE: PROCESS AND PRACTICE

Productivity

Agricultural systems are human-made communities of plants and animals, interacting with soils and climate. Unlike self-perpetuating natural systems, they are inherently unstable. Cultivation and harvesting and the biological simplicity of a few species inhibit the recycling of nutrients and makes them susceptible to attack from pests. The degree to which they can be stabilized depends on factors such as soil fertility, the extent to which animal nutrients are recycled and the diversity of the plant and animal species under cultivation. Traditional mixed farming practice, while varying widely in the type of cultivation and rapidly becoming extinct in most industrialized nations, has maintained a certain degree of ecological balance. The enclosed field systems of European agriculture relied on nutrient input from farm animals, crop rotation and variety and the natural communities of hedges and woodlands to offset nutrient loss due to the harvest and the depredations of pests. Energy inputs to the system before the age of fossil fuels were limited to horses and men to pull the ploughs, sow the seed and thresh the grain. Fossil fuels provided the fundamental breakthrough. The rubber-tyred tractor replaced human and animal labour. Chemical fertilizers and pest control agents replaced earlier and necessary biological methods of maintaining stability. Fossil fuels enabled agriculture to increase its efficiency, size and productivity, while decreasing its labour inputs.

The technologies that make this growth possible depend on large land areas and few farms for an efficient operation, a situation that has characterized farming in the industrialized countries. Relatively few, but highly productive, plant hybrids have replaced the more diverse but less productive plants of earlier farming. Worldwide, about 80 per cent of human food supplies are dependent on eleven plant species.[2]

Larger farms increasingly specialize in either crops or livestock and both are genetically engineered for maximum yield and performance. Animal feed-lots and battery poultry have intensified farm output, converting feed into meat and eggs. New breeds of mechanical harvesters have been developed for harvesting genetically modified species of grains, fruits and vegetables with minimum labour. An entire agricultural industry has evolved of which farming is only a part. What is commonly known as 'agribusiness' is based on three major components:

- an input processing industry that produces seed, machines, fertilizers, fuel and related products required for large-scale farming;
- the farm itself;
- the food processing industry which transports farm products, processes food, markets and distributes products to wholesale and retail outlets.

Thus modern farming is dependent on fuel energy not only for growing food, but for processing and distribution. Over the last century agriculture has, in effect, evolved from a labour-intensive, low-energy, small-scale and mixed farming operation to a vast industry that is capital- and energy-intensive and requires fewer and fewer people. But the benefits of high production – being able to feed more people cheaply and less grinding labour for the farmer – also bring considerable costs, in environmental and social terms, for the countryside and for the city.

Environmental costs

The first and most basic issue is non-renewable energy. Continued expansion of production, subsidized by an increasingly costly and diminishing resource, is clearly not sustainable. In the 1970s Barry Commoner pointed to the law of diminishing returns where, as cultivation becomes more intensive, greater amounts of energy subsidy must be used to obtain diminishing increments in yield.[3] The cyclical flow of energy through natural systems is simplified in industrial agriculture, which is sustained by non-renewable energy at high environmental cost. High concentrations of fertilizers and chemicals used to maximize homogeneous crops threaten soil life and deplete the humus needed for the maintenance of biological health. Streams, rivers and groundwater receive nutrients and chemicals that lead to water pollution, the destruction of aquatic species and threaten human health. Heavy machinery and tilling contribute to soil compaction and erosion and consequently the reduction of its fertility.

The complementary benefits to the soil of an animal/crop relationship disappear when crops and livestock become separate industries. Concentrating animals into feed-lots involves energy, the disposal of wastes and chemical agents to prevent disease in constricted spaces. This type of industrialized agriculture has been shown to have inbuilt

Table 5.1 *Energy efficiency of Californian crops*

Crops	Crop energy/ input energy
Field crops (barley, corn, rice, sorghum, wheat)	3.90
Raw vegetables	0.77
Raw fruits	0.54
Average of all raw foods	1.36
Canned vegetables	0.25
Canned fruits	0.25
Frozen vegetables	0.22
Dried fruit and nuts	0.63
Average of all processed foods	0.47

Source: The New Alchemist Institute, 'Modern Agriculture: A Wasteland Technology', *Journal of the New Alchemists*, 1974

inefficiencies.[4] Feed must be transported to the site. Enormous quantities of manure produced by the animals (estimated at 18 to 32 kilograms per day per animal) are often uneconomical to transport back to the fields and must, therefore, be disposed of. Once fattened, the animals have to be transported long distances to urban markets.

The replacement of manpower by industrialized food production decreases the energy value of various food crops. Figures for the energy efficiency of different crops grown in California have shown the ratios of crop energy over input energy illustrated in Table 5.1.[5] It will be apparent that despite high yields, generated by industrialized agricultural systems, there does not appear to be a viable net energy return to society. That is, the amount of energy that goes into growing, shipping, packaging and marketing the food we eat is greater that the energy we get out of it. Research conducted on farmland in upland areas of Britain concluded that as farms get larger they tend to produce *less* food per hectare on average rather than more.[6] This suggests that policies aimed at amalgamating smaller farms into bigger ones in the interests of efficiency and increased production may, in fact, be counterproductive.

Social costs

Biological stability and the sustainability of agriculture are directly related to social issues. The development of modern agriculture has been accompanied by a progressive replacement of human labour with a capital-intensive agricultural industry. The labour-intensive farming of an earlier time maintained viable rural populations. But the

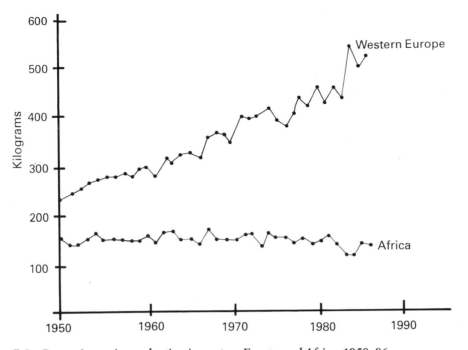

Figure 5.1 *Per capita grain production in western Europe and Africa, 1950–86*

Source: *Lester R. Brown and Jodi L. Jacobson,* Our Demographically Divided World, *Worldwatch Paper 74, Washington DC: Worldwatch Institute, December 1986*

relatively low-cost technologies, fertilizers, chemicals and machinery of modern farming have become an effective substitute for labour. An increasing number of people have moved from the rural areas to the cities in search of urban jobs. The reasons for declining numbers have been credited to increased competition, more capital to run economic operations, amalgamation of farms into large units, the centralization of agriculture, increased mechanization needing fewer people to manage more land and influences of worldwide trading blocs. These factors, however, work against the maintenance of economically healthy rural communities. Many once flourishing rural areas have been brought to economic ruin and have been unable to support or house their remaining population.

The steady disappearance of farmland in the face of urban growth has decreased the capacity of the rural areas to supply their local urban regions. City dwellers rely largely on the major food producing regions of the world for their food. A glance at the origins of the produce displayed in the supermarkets of most western cities tells us much of the distances (often many thousands of kilometres) that fruit, vegetables and meat must travel, refrigerated, before the produce reaches the buyer. In Ontario between 1981 and 1986, 4,145 hectares of these rural lands were converted from agricultural use to meet the needs of the province's expanding populations.[7] Many other regions are undergoing similar development pressures and the consequent loss of farmland. In the face of such land losses and rising energy and food costs over time, the reliance on imports from distant sources becomes an increasingly short-sighted method of feeding people. This is particularly evident as a global problem. Since 1973 the global economy is reported to have experienced an economic slow-down associated with the loss of momentum in agriculture. Data from the Population Reference Bureau illustrate the strong relationship between rapid population growth and declining capita grain production in many regions.[8] Figure 5.1 illustrates how different population growth rates are driving grain production trends in opposite directions. This can be seen by comparing trends in western Europe, the region with slowest population growth, and in Africa, which has the fastest population growth.

URBAN PROCESSES

Problems and perceptions

The problems that confront rural areas are urban based. Rural occupations, such as fruit-growing and dairy farming, which once were common in cities, are largely a thing of the past. As the cities grow larger, the vast majority of people have ceased to have any knowledge of rural values and skills. For urban people the countryside is seen as an urban playground, a recreational resource of fresh air and peaceful scenery, not as a working environment for producing food. Many have returned to the countryside to live or retire, others to find a weekend cottage; many more to camp, hike, water-ski or sightsee on holidays. Urban activities in the countryside are essentially exploitive and incompatible in terms of their environmental and social effects. For the farmer, the holiday-maker with his dog, camper and urban values is a potential menace, interfering with crops, setting fires, chasing cattle and leaving hazardous garbage that can damage property and livestock. The expanding urban edge of cities brings urban values into confrontation with rural ones, as the new urbanites complain about farm smells and rural activities. Separated as we are from the sources and processes of our food supply, and the culture that goes with it, we carry around a burden of expectations and ignorance that may be often as destructive as it may be well meaning. Cats and dogs are permitted to proliferate in the cities and are cherished, but animals kept for food are regarded with suspicion and banned by local by-laws. Two official recommendations to government pertaining to livestock keeping in cities make an interesting comparison of attitudes: the first, during

Figure 5.2 *There is a need to bring rural and urban values together so that the countryside can continue to function ecologically and as a working environment while being used for recreation*

the early years of the Second World War in England, reads as follows:

> [It is recommended] that provision should be made ... for town dwellers to keep pigs and poultry and in general to continue those rural occupations which have proved to have social, economic, and educational advantage in times of war.[9]

The second, in the city of Toronto in 1981, reads in part:

> a farm animal may be described as any animal commonly or historically used in the production of food.... Almost by nature of their purpose these live animals and poultry require extensive space, constant supervision and present obvious problems with regard to defecation and perhaps noise.... The typical city home and lot does not provide anything approaching an appropriate setting for such animals. It is proposed, therefore, that ownership of live animals and poultry within the city be banned.[10]

The meat bought at the store was once livestock, but few know of the preparation processes that precede the packaged product. We have little idea of how to handle or manage the animals that are the source of our food and impose sentimental value on 'fluffy baby animals', which inhibits a straightforward acceptance of their use as farm livestock. The slaughter of farm animals for food is an unknown and unwanted experience. It is clear that there is a need for an awareness of where our food comes from for urban children and adults alike.

The city presents biological problems that cannot be ignored without at the same time sterilizing the minds and habits of its inhabitants. Our perceptions of the urban environment and the things we do there have, as I showed in Chapter 1, been moulded by the isolation of modern urban life from what were once rural values, but which still remain one of the fundamental elements of life support in cities. The preoccupation with leisure and aesthetic values as the exclusive function of urban parks and spaces, and the inherent assumption of their non-productivity have remained since the nineteenth century, the unquestioned dicta of parks departments, designers and the public.

Redressing the balance

The 1990s has brought a heightened public awareness that diverse and productive cities are a fundamental basis for a sustainable future. Soil conservation, modern versions of traditional small-scale farming, the health problems inherent in chemical food production and concern for a greater control over personal and community destinies are being expressed in a groundswell of organizations that are searching for a more humane and integrated relationship with the natural processes that sustain life. Many private and commercial operations have taken to the land to produce 'organically' grown food uncontaminated by chemical additives or preserved in tin cans. There is a search for greater self-reliance, a more direct connection with land and the satisfaction of having more control over personal diet and health. So the question must be asked: how are urban issues relevant to the problems of food and farming?

An answer may lie in the fact that, as I have previously suggested, rural problems are urban based. There are land issues here that affect not only western cities but, even more critically, cities in the developing world.

If we look back in history, its becomes obvious that the disconnection between urban people and the land is a contemporary malaise. In pre-industrial cities, necessity dictated a measure of integration between urban and rural occupations. Mumford notes that a good part of the population of medieval cities had private gardens and practised rural occupations within the city. In addition, burghers had orchards and vineyards in the suburbs and kept cows and sheep on the common fields under the care of a municipal herdsman.[11] Pigs and chickens were kept, and, in the early days, these acted as town scavengers. Until provisions for street cleaning were effected, Mumford notes, the pig was 'an active member of the local Board of Health'.[12]

The New England towns of the United States maintained a similar balance between rural and urban occupations up to the end of the nineteenth century. In Britain, prior to the First World War, a thriving dairy industry existed in Liverpool. Many families from the Yorkshire Dales went into the business of supplying milk to a fast-growing urban population, keeping cows in the city and selling the milk to homes on the city streets. The story of these enterprising families is told through one family who stayed in Liverpool for twenty years before retiring to a farm in the Yorkshire Dales that they had bought with the profits of their Liverpool milk business. The business started with two cows and ended with forty-six cows and three horses. The Liverpool house where they lived was at the end of a row, and the front room formed a dairy where people arrived at all hours to buy milk. A short distance away were the buildings housing the cattle. The daughter took a horse-drawn lorry to the local cemetery in the summers to collect loads of grass cuttings which had been put into piles by the cemetery maintenance men. Hay for the animals was obtained from a local farmer who would dump a load on his way to the city market and load up with manure on his way home.[13]

Urban food growing by necessity

While this Yorkshire Dales initiative is an elegant example of self-sufficiency in an industrialized society, the absolute need to be self-sufficient is all too clearly demonstrated in many developing countries. Population growth, and the legacy of colonialism that resulted in land monopolies and export-based agriculture, have left many peasants landless and poverty stricken. For example, 2 per cent of landowners in El Salvador hold 60 per cent of the land, while almost two-thirds of rural families have little or none.[14] White farmers, who comprise 1 per cent of Zimbabwe's people, control 39 per cent of all land.[15] Urban food growing is, therefore, one of necessity and survival for the rural people who stream into the cities. A large number of these 'rural refugees' have the survival skills to produce food on their own given access to the necessary resources. It has also been noted that these survival skills could also be used to help such people adapt to urban life, since urban agriculture contributes to increasing social contacts.[16] The economic value of garden plots, where space is available, is particularly important in developing countries where the food budget accounts for a much higher proportion of total family expenditures. These are often 50 per cent, reaching 70 per cent for urban families in India, compared to 25 to 30 per cent for the average American family.[17] Sachs notes that the potential contribution of urban agriculture to the food supply of the urban poor, based on a 200 square metre garden over a six-month growing season, would provide one-fifth of the optimum food intake for a family of five, or a great deal more in terms of present low nutrition standards. Estimates have shown, for instance, that in 1980 340 million people in eighty-seven developing countries were not getting enough calories to prevent stunted growth and serious health risks – a 14 per cent increase over 1970. More recent World Bank estimates have shown that 940 million people still consume too few calories to support an active working life.[18]

The pursuit of rural occupations in many cities, therefore, clearly demonstrates that it has largely been based on the imperative of necessity. The same is true, therefore, of war and emergency. During the German occupation of Denmark in the Second World War, it was the food grown in gardens that saved the citizens from starvation.[19] In Britain, efforts to make the most of limited resources led to the setting up of the Pig Keeping Council in 1939 by the Ministry of Agriculture. Its original objective was to move pigs from the farms, which were threatened by shortages of imported feedstuffs, to the villages to encourage the use of household wastes for feed. These efforts, however, were soon reflected in an *urban* livestock movement. Since the bulk of edible waste came from the cities, pig- and poultry-keeping naturally evolved as a major urban activity. Pig-keeping spread on to bombed sites, in back streets and allotments and included policemen, firemen and factory workers among the devotees. There was a pig club in London's Hyde Park and another within 180 metres of Oxford Circus. In 1940, in response to the increasing demand for eggs, the Domestic Food Producers Council set up a Poultry Committee which recommended households be encouraged to keep poultry for egg production. The backyard food production movement became officially accepted with an Order in Council which suspended restrictions on the keeping of pigs, chickens and rabbits, subject to certain public health requirements.

By 1943, there were 4,000 pig clubs comprising some 110,000 members keeping 105,000 pigs. While the promoters of the pig-keeping movement originally intended it to be a rural activity, it became urban because the food wastes of modern society were mainly in urban areas. Prior to this, the disposal of waste food levied a heavy charge on public rates. Many of the clubs were organized into co-operatives, livestock being fattened collectively, rather than by individuals, lending greater efficiency and organization to the activity.[20] By 1942 there were 916,000 registered poultry-keepers keeping one to twelve chickens and nearly 264,000 with thirteen to thirty chickens with a total poultry production of some 16 million. Domestic rabbit clubs were formed, and by 1943 amounted to 2,700 individuals keeping 252,000 breeding does. Other livestock activities included bee-keeping and goats. In England and Wales at that time there were about 30,000 bee-keepers controlling some 429,000 colonies of bees. While bee-keeping, by its very nature, is not an urban pursuit, there were many bee-keeping operations that were distinctly urban or semi-urban in character. One of the most progressive associations of

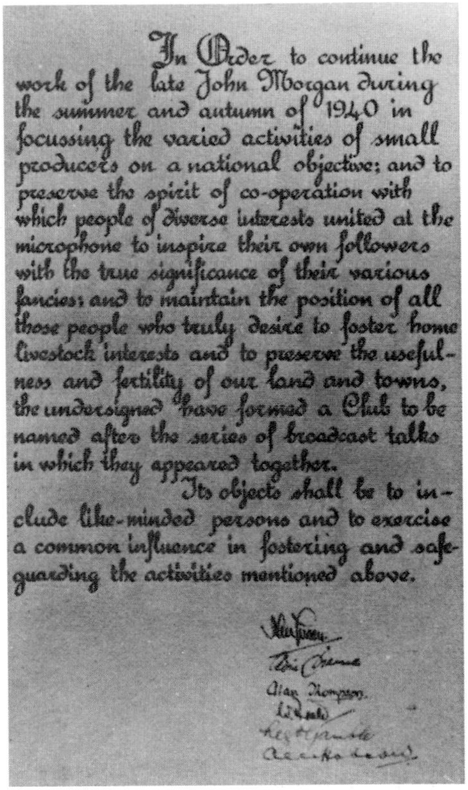

In Order to continue the work of the late John Morgan during the summer and autumn of 1940 in focussing the varied activities of small producers on a national objective; and to preserve the spirit of co-operation with which people of diverse interests united at the microphone to inspire their own followers with the true significance of their various fancies; and to maintain the position of all those people who truly desire to foster home livestock interests and to preserve the useful-ness and fertility of our land and towns, the undersigned have formed a Club to be named after the series of broadcast talks in which they appeared together.

Its objects shall be to in-clude like-minded persons and to exercise a common influence in fostering and safe-guarding the activities mentioned above.

Figure 5.3 *The Charter of the 'Back to the Land' Club, formed early in the Second World War by John Green, who was in charge of farming and gardening broadcasts at the BBC in London*

The Club provided a focus for the various organizations concerned with food production, including the Pig Clubs, Domestic Poultry Keepers, the National Allotment Society and others. The Club also served as a technical forum and defensive organization to solve problems such as urban by-laws, treatment of wastes, administration of rationing, livestock in schools and so on. The Club lasted twenty-six years, until rationing was discontinued in Britain

bee-keepers was in the city of Birmingham, where facilities for an instructional apiary were provided in one of the public parks. It is interesting to note that the planning of parks and gardens was of particular concern to bee-keepers, since the selection of plants in the city's parks would have considerable impact on nectar supply. The maintenance of plant diversity that was discussed in Chapter 3 can be seen to be a crucial factor in a functional productive urban landscape. In addition, the value of bees as pollinators for allotment and backyard gardens, is, in the opinion of modern bee-keepers, much more important than their value as producers of honey. Goat-keeping was practised intensively partly due to the shortage of fresh milk and partly because the waste on which goats thrive meant that they were not competing for food with other animals. Milk yields averaged over 4.5 litres a day throughout the year from goats that were entirely stall fed and with a yard of some 17 square metres for exercise.

The 'Dig for Victory' campaign during the Second World War in Britain and other countries showed that the production of fruit and vegetables in or near cities could have a significant influence on food production when people are in need. Production from allotment and garden plots in Britain reached a peak when the number of allotments almost doubled from a pre-war figure of 740,000 to 1.4 million. In a parliamentary debate reported in Hansard in 1944 it was estimated that 10 per cent of the food grown in Britain came from this source.[21] During the years when urban food production was at its height, agricultural shows took place in a variety of places, including the basement of John Lewis' shop in Oxford Street, in industrial areas and on football fields. John Green, founder of the 'Back to the Land' Club and in charge of BBC broadcasts on farming and gardening at the time, recorded a visit to Bethnal Green after one of the worst air raids of the war during which he saw the people feeding their ducks on the canals and visited a poultry show at the Working Men's Institute. The inner city had just acquired a new unity with the countryside.[22] In Canada, Vancouver citizens urged by the government to 'plant a wartime garden', and aided by a city decision to rent vacant land at a nominal fee, produced some 31,500 tonnes of fresh vegetables and fruit in 1943. This was equivalent to Canadian $20 million worth of supermarket produce at 1979 prices.[23]

As these examples illustrate, crises and shortages lead to the adoption of alternative strategies for survival. Since the end of the Second World War, however, an affluent urban society has not had to concern itself with these basic needs as food has become readily available regardless of season or distance. But with the shift of societal values towards a better relationship with nature, an interest in 'organic' farming, human health and diet has emerged. One of the indicators of these changing values is the rebirth of the farmers' market. Although they had never completely disappeared, pressures from the supermarkets and large-scale farming during the 1950s and 1970s reduced their competitiveness, and suburban expansion replaced the farmland close to the cities that had been the mainstay of their economy. Since the mid-1970s, farmers' markets have revived and by 1993 were reported to number some 2,000 in the US.[24] Several forces are reported to have led to this revival.

The first has to do with small-scale economics. The uncertain financial returns from selling products to wholesalers has forced farmers to sell direct to the customer. The farmers' market provides an ideal outlet, one that has kept many small farmers in business. The second concerns changing consumer attitudes which have also had an influence. An increasingly sophisticated public is looking for variety, taste and pesticide-free produce that the 'standard supermarket lettuce wrapped in cellophane and one type-fits-all green peppers' cannot match.[25] And in a planning context, urban planners have discovered that farmers' markets are an effective way of revitalizing downtowns.[26]

Thus the opportunities for re-establishing constructive links with the land are tied to the food we eat, and it is in the cities that these links must be re-established. They are the places where alternatives to destructive technologies can be explored and demonstrated through direct experience with soil productivity, the recycling of nutrient and material resources and urban metabolic processes. We can learn about energy and food production, farming practices, market gardening techniques, rural affairs as well as urban ones, while turning 'waste' into useful products. To do so requires that farming at appropriate levels becomes an integral part of the city's open space functions. It also requires that the overall physical and social structure of parks and open spaces must be reshaped when their essential productive value is recognized. The opportunities for doing so should, therefore, be explored.

RESOURCES AND OPPORTUNITIES

Some issues in the developing nations

It is apparent that cities have significant potential for small-scale agriculture, so it is here that the task of building a richer and more productive city environment can be achieved. This fact is borne out in cities in the developing nations that must produce food for large populations, often with limited space and energy for transportation, and minimal financial resources to import food. In China, government policy has aimed to create producer rather than consumer cities. At least 85 per cent of the vegetables consumed by urban residents are produced within urban municipalities.[27] Shanghai and Bejing are self-sufficient in vegetables, and many Chinese cities also produce large quantities of poultry and pigs and other essential foods on the edges of towns and cities from where it can be transported at minimum cost to the population centres.[28] Food production in urban Kenya is of widespread economic importance to the survival of many Africans, and in 1985 cities produced an estimated total of 25.2 million kilograms of crops worth US$4 million.[29]

A key factor in the developing nations, where poverty is a major problem, is the need to link land, food production and waste recycling with employment and income. Many Asian countries, such as Indonesia, practice urban aquaculture where fish ponds provide the opportunity for fish and rice production to be integrated with animal and human waste treatment and absorption. For over 2,000 years Chinese fish ponds have produced fish fed from grass clippings with animal and human manure feeding the ponds. Fish farmed from these ponds account for 2 million tonnes or just under 20 per cent of total fish production in China.[30] In India, thousands of poor acquire their fuel, and supplement their incomes, by gathering animal dung to make fuel patties.[31] In some cities like New Delhi, hotel and restaurant wastes are auctioned off to poultry farmers.[32] The exploitation of urban organic wastes and sewage for food production are to be found in the eastern fringe and hinterland of Calcutta. Sewage nourishes fish ponds and paddy fields; productive vegetable farms are located on the garbage dumps created by the city in the 1860s, and use the natural compost generated by them. The practice of sewage farming, the growing of grass and other crops on municipal land with liquid wastes, was brought from England to India in the nineteenth century and has survived in some twenty-five cities and towns.[33] This practice can also be seen in other Asian cities such as Jakarta, Indonesia.

The disposal of garbage is a universal problem of expanding cities worldwide, and is often linked to recycling and income by scavenger groups. In 1992 an article in the *Jakarta Post* praised the scavengers of that city for collecting and sorting the garbage, some 6,000 tonnes per day, and recycling it for a living. Garbage pickers at the garbage sites take their finds to nearby junk dealers who buy their old cans, plastic materials and bottles and sort them out for further recycling.[34] Recyclable plastic, after being cleaned, is in demand by plastic factories where it is reprocessed and used again. Similarly, iron scraps can also be sold for recasting after being melted down. Other materials include used cans that are in high demand by kerosene stove-makers; soy ketchup bottles have a good market value since they can be refilled; furniture shops are grateful for new supplies of crates since it is cheaper to use parts of crates than to buy new ones.[35] Some industries such as shoe-making factories give their scrap leather to small businesses that recycle it into new products.[36]

A three-year pilot project to introduce a self-financing Integrated Resource Recovery system in Bandung (the second largest city in Indonesia), reveals the close relationship between environmental enhancement, social organization and job creation and income. The research identified two contradictory waste management systems at work side by side in Indonesia: one that is typical of other developing countries – the 'formal' system operated by local government based on the western-style concept of collection-transport-dumping of waste – and its 'informal' counterpart – the scavengers who collect materials with a resale value and sell it to industries, which in turn recycle it into reusable products. As the volume of waste increases in the former, so do costs and environmental health concerns, which then require the introduction of increasingly sophisticated technologies.[37] A participatory action research project began in which ways were sought to involve a group of families and scavengers, break down social barriers and develop a sense of common community purpose in the project, and establish a centre for organizing common action. Within three months, the initial families who had been trying to survive individually, transformed themselves into a dynamic and creative community. By the end of the project in 1986 the community had grown to eighty-eight families earning a living by sorting and recycling recovered paper, glass and metal, composting organic wastes for sale and intensive urban farming, raising rabbits, improving their housing and other community activities.[38]

This experience became the basis for developing an Integrated Resource Recovery module as the building block of a dispersed system of waste processing. It is one that can be developed incrementally as market opportunities permit, and as a socially and

Plate 5.1 *Farming on composted garbage in Jakarta, Indonesia*

Table 5.2 *Scavenging Bandung*

Type	% or number	Value (Rp)	Recycled into
Paper	38.9%	70–100/kg	paper, cardboard,
Metals	22.0%	40–300/kg	kitchen utensils, various
Bottles	20.6%	75/piece	glassware, hoes, caps for
Textiles	12%	200/kg	bottles, cattlefeed, shoes,
Drums (steel)	100/month	150–800/kg	soles, etc.
Drums (plastic)	100/month	5,000/piece	
Sacks	26,000/month	1,500/piece	
Tyres	90/month	125/kg or 500–850/piece	

Notes: In any one district, or *kecamatan*, there are on average one to fourteen *lapaks* (leaders of scavenger communities). Each *lapak* usually employs around eleven to sixty scavengers, with the total number of scavengers in Bandung estimated at about 2,000 to 3,000. Their level of education varies, with 53 per cent having reached primary school level, 25 per cent secondary school level and 22 per cent high school level; these figures include drop-outs. The daily income of a *lapak* is between Rp 5,000 and Rp 5,000,000, that for a scavenger between Rp 1,000 and Rp 5,000. The total investment for waste collection in Bandung (recorded) is Rp 408,490,000. The table shows the volume and type of waste collected for recycling, and where it is sold

Source: Hasan Poerbo, 'Urban Solid Waste Management in Bandung: Towards an Integrated Resource Recovery System', *Environment and Urbanization: Rethinking Local Government – Views from the Third World*, vol. 3, no. 1, April 1991

environmentally viable alternative to conventional, large-scale waste management. Of key importance, it serves as an alternative to dumping unwanted materials into the local rivers, an endemic environmental problem in Indonesia. In effect, the two approaches – the formal and the informal – represent two different views of waste. The former considers it as a health and environmental hazard; the latter considers it as an economic resource from which marketable products can be derived. This latter approach thus achieves a number of objectives: it reduces volumes to be dumped; it reduces the need for financing and subsidizing waste management; it creates jobs and income opportunities; it creates social and community cohesion; and it has significant environmental benefits. As Table 5.2 shows, the economic, social and environmental gains of such a model can be considerable, when it is developed as an integrated approach to the urban scene.

Cities as multi-cultural places

These examples of how collection, recycling, food production and employment are linked to the city's unwanted products are by no means limited to cities in the developing world. What is a universal condition demanding ecological, social and economic solutions for Jakarta, or Delhi, is also one for New York and most other major western cities. For instance, the non-profit organization known as 'WeCan', founded in 1986,

Table 5.3 *Countries of birth of immigrants to Canada*

Top 10 countries in 1991		*Immigrants who came to Canada between 1981 and 1991*	
United Kingdom	717,745	Hong Kong	96,540
Italy	351,620	Poland	77,455
United States	259,075	China	75,840
Poland	184,695	India	73,105
Germany	180,525	United Kingdom	71,365
India	173,670	Vietnam	69,520
Portugal	161,180	Philippines	64,290
China	157,406	United States	55,415
Hong Kong	152,455	Portugal	35,440
Netherlands	129,615	Lebanon	34,065
Total number of immigrants	4,342,890	Total number of immigrants	1,238,455

Note: Of the total number of immigrants living in Canada (4,342,890), most have come from the United Kingdom. But trends – and the numbers – are changing

Source: Statistics Canada

acts as a financial exchange for New York's enterprising homeless who collect cans and bottles and sell them to WeCan for a 5-cent refund, who then sell them to manufacturers for 6.5 cents per container.[39] The status of western cities as multi-cultural environments has become a fact of life. According to Statistics Canada, the origins of immigrants to Canada are becoming increasingly diverse, even though the percentages of Canadians who are immigrants has remained the same since the Second World War. The rapid changes in immigration patterns are illustrated in Table 5.3. It shows that while there are more people of British birth living in Canada, between 1981 and 1991 there were more immigrants from Hong Kong, Poland, China and India than from the UK. In 1961, 90 per cent of immigrants came from European countries; between 1981 and 1991 this figure had dropped to 25 per cent. The impact on this trend can be seen in big cities such as Toronto, which has the highest proportion of immigrants of any metropolis, and where 38 per cent of the population is immigrant.[40]

Multi-culturalism is reshaping the physical and cultural character of modern cities. The traditions that ethnic groups bring with them have always enriched the character of inner-city neighbourhoods, with their street markets, residential areas, small industries and restaurants. The productive urban landscapes of mini urban farms and micro vineyards are the hallmarks of Portuguese, Chinese and Italian neighbourhoods. This vernacular is hidden away in alleys, rooftops and backyards in neighbourhoods all over the city. Here one finds an extraordinarily rich variety of flourishing gardens, houses, streets, people, expressing ideas of personal territory, particular ways of using space, what is valued and a rich urban tradition. Behind the front and backyard garden fence one will find productive vegetable gardens, grapevine covered trellises provide summer shade and grapes for autumn wine-making. These vernacular landscapes reflect still surviving rural skills and cultural connections with the land, and represent very different views of the city in both function and aesthetic priorities.

The most dramatic trends are, however, in the changing social and physical make-up of older suburban areas. Once the domain of Anglo-Saxon, white-collar workers, many of them are experiencing a significant influx of diverse ethnic groups. The urban agriculture of the inner city is moving with them to transform hitherto conventionally landscaped and sterile suburban backyards. Mini farms of small livestock, vegetable and fruit gardens are taking over the previous non-productive lawns. What these trends show is that such changes are driven by the special needs of different groups. The necessities of change usurp what planning by-laws permit. For instance, the richness and mixed land uses of many a downtown ethnic neighbourhood is often prohibited in the outlying suburbs of adjacent municipalities that set rigid restrictions on what may or may not be done to individual houses or properties. Exclusionary land use controls are ostensibly based on concerns about exceeding current infrastructure capacity, or damaging the environment. But many such regulatory barriers actually stem from communities' efforts

Plate 5.2 *Food production by necessity and tradition in a North American city*

Every available space is cultivated and used to the maximum in many ethnic areas. Front and backyards supply vegetables for the family – some families being almost self-sufficient. Flowers, religious icons and decoration fill ground plots and rooftops; small cafes and backyards are sheltered from the sun by grape vines that are grown for wine-making. This rich urban tradition is a far cry from the standard park provided by the municipality. Installed and maintained at public expense, it has little diversity and no productive uses

a *Backyards sheltered by grape vines, which provide wine and shade*

b *Gardens at ground level with grape vines shading the house, and vegetables growing in the front yards*

c *The market catering to a highly diverse cultural community*

d *The municipal park*

to keep out certain groups of people. Local zoning in the US, for instance, frequently reflects community elitism or racial ethnic prejudice.[41]

As we will see later in this chapter in our discussion of the Federation of City Farms in England (pp. 238–43), there is a growing recognition that parks systems must become meaningful for poorer neighbourhoods and provide links between urban communities and the land. In the US, the San Francisco League of Urban Gardeners, a grass-roots non-profit organization, has recognized the realities of North American multi-cultural society, poverty and unemployment, and is committed to building gardens in culturally diverse neighbourhoods. With 40 per cent of the youth under 25 unemployed, the organization functions as a job training programme, employing youths as trainee carpenters, landscape construction workers, tree planters and related skills.[42] The Uhuru gardens to be built in south-central Los Angeles on a lot still covered with rubble from the Watts riots of 1965 is another example of a project that will demonstrate a restoration of part of the city, by reconnecting people with the land, and by vocational training for jobs in the green industries: 'Uhuru Gardens would marry an environmental ethic to landscape training and internships for community residents.'[43]

Stan Jones, a former landscape architect for the San Francisco League of Urban Gardeners has noted that '[Multi-culturalism] is not a mere acceptance of cultural differences; at best, it actually celebrates those differences – such as when gardeners share produce and learn to eat and even cultivate exotic vegetables'.[44] This comment also

Plate 5.3 *The land resources in the city*

While these are the consequence of formless urban growth, they can also be seen as an opportunity for the future. They include electrical transmission lines, highway intersections and interchanges and rooftops

points out another reality, that different cultural groups use space differently and have different behaviours and needs. Jones observes that

> in Chinatown, you'll see parks that have a certain number of linear feet of seating and other features following William Whyte's guidelines for urban spaces. But segregated in a quieter corner you'll see groups of older Asian women.... They don't want to be right up front, they don't want to be near the street, they don't want to be looking at the action – contrary to everything Whyte said.[45]

The tendency for cookie-pattern park design in multi-cultural neighbourhoods does not recognize that different physical expressions for parks evolve from understanding the cultural, psychological and behavioural needs of the people who use them. In addition, the conventional open space provided by the municipality through public taxes involves high maintenance cost, has little diversity and provides no economic returns. And as Karl Linn (the 1950s pioneer of community design), Randolph Hester and others have shown, this can only be achieved when the people most concerned participate in the design of their own places. The inevitability of the changing urban landscape requires an ecological response that recognizes the fundamentally similar forces of natural and human communities, and helps them develop in their own ways and with their own imperatives and aesthetics. It is necessary to rethink the basic nature and functions of how urban space is used.

Physical and energy resources

It is plain that modern urban tradition has been moulded by a combination of economics, technological and aesthetic influences and values. This is reflected in the fact that vast amounts of nutrient energy and land are conceived of as waste. The idea that the city itself should be required to contribute to the production of such basic essentials as food involves, as we saw in the Indonesian example (p. 216), a shift in social and ecological values. From this perspective the urban environment may be seen as a generator of vast *resources* in nutrient energy and land. The food industry and the leaf litter collected by the parks department, for instance, provide potential resources for enriching soils for gardens, parks and rehabilitation sites. Wasted land created by urban expansion, planning regulations and inefficient single use zoning may, in fact, be regarded as an invaluable opportunity for the future when necessity may well dictate its more productive and intelligent use. The city's spaces, in effect, will be seen as having value beyond the recreational and aesthetic purposes generally ascribed to them.

Land

The availability of urban land is potentially enormous in almost all major western cities. Railway, public works and public utility properties, vacant lots, cemeteries and industrial lands form a major proportion of unbuilt-on land that has been, and remains, sterilized or ineffectively used. The amount and type of land naturally varies depending on the peculiar political, economic and urban renewal conditions of each place. But they are reminders of the vast areas that exist as cities expand on to agricultural areas, and as more and more land lies vacant and fragmented within them.

Residential property

At the smallest personal scale, the front or backyard provides some of the best opportunities for food-growing in terms of energy, efficiency and direct benefit. As I showed on p. 206, commercially grown and processed vegetables and fruit are the most energy-intensive of all crops: it requires more energy to produce them than the energy benefits they return. A comparison between the yearly energy budget of a residential lawn, 20 square metres in area, and the use of the same space for productive crops has revealed some interesting results. Each unit of energy invested to maintain the lawn, in human work, fuel for the mower, fertilizer and pesticides, returned 6 units of energy as lawn clippings. If the clippings were discarded, the net production efficiency of the lawn could be described as zero. With the same area planted with alfalfa, each unit of energy invested returned 22 units in crop production. In addition, the alfalfa produced enough digestible nutrient per square metre to support the production of 450 grams of rabbit meat (Figure 5.4).[46]

With respect to diet, the food produced on a residential lot is a small-scale, human energy-intensive operation. For some less affluent levels of society, growing food has definite economic advantages as the price of food continues to rise. For others it affords opportunities for productive spare-time work and pesticide-free and tasty vegetables. An experiment designed to determine the economics of food-growing on the front yard of a small residential lot of 30 square metres was undertaken in the 1980s by the author.[47] Whether the gardener makes money or loses it on the home vegetable plot had seldom been investigated. Total returns for the vegetable patch in the first year were Canadian $136.43. Start-up costs were $159.49, exceeding returns by $23.06. The second year produced total returns of $230.40 with costs of growing seedlings and soil additives being $56.63. Thus a net gain in produce was realized of $173.77. Calculated in returns per square metre the garden produced $7.68. Four years later this had increased to about $10 a square metre.[48] Analysis of productivity was based on weekly evaluations of the cost of produce at the local market and then applied to produce gathered from the

Yearly energy budget of a lawn compared to an alfalfa patch

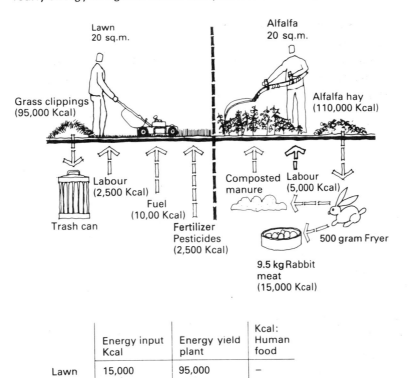

	Energy input Kcal	Energy yield plant	Kcal: Human food
Lawn	15,000	95,000	–
Alfalfa	5,000	110,000	15,000

Figure 5.4 *A comparison of two equivalent sized residential yards of 20 square metres, the one producing lawn grass, the other alfalfa*

For every unit of energy invested (human labour only) in the production of alfalfa, 22 units of energy were returned in crop production. In the case of the lawn, the net production efficiency (assuming the grass clippings are discarded) is zero. It has been calculated that the rate of energy use for the maintenance of some 6.5 million hectares of lawns in America exceeds the rate for the commercial production of corn on an equivalent amount of soil

Note: *All notations represent annual totals*

Source: *The Farallones Institute,* The Integral Urban House, Self-Reliant Living in the City, *San Francisco: Sierra Club Books, 1978*

garden. The cost of labour developing and maintaining the garden was not included. Thus it can be seen that even a tiny plot of land in a city with a relatively short growing season (May to September) can produce a considerable net gain in produce, and increase the margin of self-sufficiency for low income families. The educational advantages of direct contact with plants and soil are interwoven with a need for greater self-reliance, less dependence on the centralized food systems of agriculture and pesticide-free food. While home food production is a question of personal choice, it is clear from the number of publications and organizations concerned with food in cities that there is a growing interest in the concept of self-reliance and an improved quality of life. Among the organizations concerned with urban problems was the Farallones Institute in Berkeley, California (disbanded in the 1980s), which attempted to demonstrate by practical means how life-support systems can be integrated in an urban house in such a manner as to conserve energy and resources and contribute to the health of the city's environment. Another, the Institute for Local Self-Reliance, Washington DC, provides guidance to local communities in recycling wastes, energy efficiency, urban agriculture and related issues. In Britain, there is the National Federation of City Farms (see pp. 238–42) and the Centre for Alternative Technologies in Wales. And in Toronto, Canada, various organizations include 'Gro Together', a group of volunteers who organize food drives and food banks for the urban poor.

Thus, residential resources for urban farming are considerable and include any space where the individual can grow a few tomatoes or lettuce. Private food-growing does not affect public space directly, but the city as a whole becomes richer and more productive because of it.

Wastelands and allotments

The demand for allotment gardens in or near the cities has been rapidly growing for many years, particularly among people who live in apartments in locations where outdoor space is at a premium. Increased leisure time, early retirement and unemployment are credited with the increasing demand for a plot of land to produce organically-grown food. Allotments can be fitted to any shape of property. The small-scale and labour-intensive agriculture of the allotment is ideally suited to the small or awkwardly shaped space that has little potential for many other uses. Allotment food production is more concerned with high-value crops such as tomatoes, beans, onions and lettuces that give better financial returns on time invested than crops such as potatoes and cabbages, which may be cheap and plentiful in the shops. The productivity of such a method is potentially high. It has been estimated that the standard allotment garden of 250 square metres may yield, under experienced management, 20 tonnes per 0.4 hectares or approximately 3.5 tonnes per plot per year. The Friends of the Earth calculate that this level of performance, repeated on the 101,200 hectares of land in British cities that were estimated by the Civic Trust to be lying idle in the late 1970s, would yield some 5 million tonnes of food.[49] Another study of the economic development potential of community scale gardening in Boston reported estimates that a gardener producing 225 kg of

Plate 5.4 *Flat rooftops on many industrial buildings offer great opportunities for productive uses, such as gardens, greenhouse production using waste building heat and mini wetlands for wildlife*

vegetables on a 54 square metre plot could save a minimum of US$235 at average 1979–80 market prices.[50]

Other urban wastelands include the hundreds of hectares of rooftops that for the most part lie desolate and forgotten in every city. Yet many present open space opportunities that could be turned to productive use. In Toronto, for instance, the Toronto Food Policy Council is encouraging community gardens on the apartment rooftops that exist in that city.

Heat

A prodigious amount of heat energy is pumped into the city atmosphere from heating and cooling systems and industry, and this has a major impact on urban climate (Chapter 6). This same energy must, however, be regarded as a resource rather than a problem in the context of urban farming. The wasted heat energy from buildings and generating stations may be considered in several ways. One is the question of conservation through better design and insulation, about which considerable discussion and application has taken place in the 1990s. Another is the capture of waste heat that can be made use of for other purposes. Sweden harnesses approximately one-third of the waste heat from its power plants for commercial purposes.[51] The same principle of tapping waste heat has great potential for food growing, particularly in cold countries where the growing season may be limited to four or five months of the year. The need for greater self-reliance in the future is being recognized as food, imported during the non-growing season and based on high-energy inputs, becomes more costly. This potential has applications at various levels, both at the domestic and commercial scale. Self-reliant organizations such as the New Alchemists have for years been developing practical applications of integrated living systems in the US, where greenhouse agriculture becomes part of a total biological cycle of life systems. In the urban context the heat pumped out from industrial processes may be connected directly to food-growing industries. The use of flat rooftops of many industrial buildings, using lightweight hydroponic (soilless agriculture) techniques, could begin to make more efficient use of both space and energy resources that are normally discarded. The industrial parks of many North American cities could become agricultural as well as industrial producers, combining functions that have traditionally been separated. For instance, whisky distilleries in Scotland cultivate eels (a highly prized delicacy in Europe) in large tanks using the heated waste water from the productive process.[52] The same principle could have even greater relevance to the city since the costs of transportation can be minimized, and agricultural produce can be grown in direct association with local markets.

Nutrients

A key factor in urban farming is the question of soil enrichment and management. Many urban soils have been sterilized through constant disturbance and pollution. Consequently they have little fertility. But the move in many countries to a thriving agricultural industry during the Second World War was made possible by the immense inherent wealth of nutrient resources that is available in the city. As discussed on pp. 215–17, this is true of Chinese and other Asian cities where human waste has been important to agricultural development and fish production. In the 1990s recycling and waste management has become a priority in most industrialized countries, largely due to the diminishing availability of landfill sites, and the increasing cost of transporting wastes sometimes great distances from local boundaries. Between 1980 and 1987, Philadelphia's disposal costs rose from US$20 to $90 per tonne.[53] In 1992, the Minister of the Environment, Ontario, decided not to allow southern urban municipalities to transport waste to northern communities, forcing them to find alternative solutions locally. And there is increasing citizen opposition to incineration due to concerns about mercury, lead and dioxin pollution from the stacks, and heavy metal contamination of groundwater from incinerator ash.[54] Throughout the world, however, at least one-fifth of municipal waste is composed of organic kitchen and yard wastes.[55] In addition, other urban nutrient resources are the unwanted products generated by the food industry, municipal parks and public works departments. Traditionally dumped or burned at a mounting financial cost to urban municipalities, an increasing number of cities have initiated composting stations that provide a valuable soil amendment for increasing the organic matter of poor soils, or water-conserving mulch for plants. Where sewage is free of heavy metals, or other industrial contaminants, the sludge can be composted using contemporary technology, as was initiated in 1988 in the Connecticut town of Fairfield. The compost centre mixes sludge with leaves, grass and branches, and is reported to produce an odour free compost in about twenty days.[56]

IMPLICATIONS FOR DESIGN

The basis for form

Historically speaking, the form of early towns and villages was dictated by their relationship to the agricultural fields where food was produced. The symbiotic relationship between the fields that produced food for the town, and the town that returned its refuse to enrich the fields, was necessary to ensure survival. The role of open space within the town, as Mumford has observed, was primarily a functional one; it was important to citizens for growing useful produce or for livestock.[57] The agricultural co-operative settlements (kibbutzim) in Israel show a similar pattern; a direct relationship between habitat and food-producing land. It was initially for survival in a harsh environment that the kibbutz system arose.[58] But new towns are few and far between and have little impact on the problems of existing cities. It is with the reshaping of the existing city landscape – with the productivity of its soils – that we must be concerned. The breakdown between people and direct connection with the land began as a consequence of industrialization, growth and mass migration into the cities for non-rural work – a phenomenon that continues in developing countries. The creation of urban parks in the eighteenth and nineteenth centuries were based on spiritual and leisure needs rather than the functional necessities of food-growing. This perception of open space – to provide recreational and aesthetic amenity at public expense, thereby satisfying the soul rather than the stomach – has persisted as the main objective of the parks. While recreation is today an essential facet of urban life, it is, as I have tried to show, only one of the many functions that urban space must serve as we attempt to build an ecologically and socially viable future for our cities. Urban agriculture will increasingly become a necessary urban

Plate 5.5 *Productive market gardening on industrial lands in a suburban area of North York, Metropolitan Toronto*

A productive well-tended landscape that returns food energy for energy invested. Landscape with edible plants; an example by default of a new interdependence between city and land that has been absent since the pre-industrial city. On the other side of the street, York University's immaculate lawns and non-productive landscape; perpetual investment of energy for no energy return

land use function that should be publicly supported as long as poverty, hunger, multi-cultural traditions and land-connected recreational trends continue to influence city life and city form.

As we have seen, the European allotment garden, located on the edge of the city on railway land and any other piece of ground that can be leased, is a precious resource. High demand and limited land have made it so. The garden plot is the citizen's summer cottage, often within cycling distance from home. Each plot is laid out with loving care, with a shed that includes storage for tools and often living space. On a larger scale, other examples offer fascinating insights into the potential of urban agriculture to make productive use of derelict land while providing fortuitous alternatives to the treatment of unused land.

Farming on industrial land – a commercial venture

An examination of the city's open spaces has shown that industrial lands take up very large areas, particularly on its fringes. These neglected places contribute to the city's visual blight, particularly when they become encircled by development. One such area lies in the borough of North York in Metropolitan Toronto. Once on the fringes of the city, it has seen tremendous growth over the last twenty-five years or more. The York University campus, first located here, was followed by major industrial, residential and commercial development on previous agricultural land. The industrial area adjacent to the university, owned by four oil companies, totals 91 hectares, about half of which have been developed as tank farms and related plant. The remaining lands have not been built on, the consequence of hard economic times for the oil industries. Visitors to this part of the city are greeted by two landscapes as they drive north. On one side is the university laid out with immense spaces of mown turf and street trees lining the road, the product of conventional design vocabulary. On the other are the oil company lands, part of which have been transformed into fields of corn, tomatoes, zucchini (courgettes) and peas. These lands, surrounded on all sides by industrial and residential development, were in the early 1980s leased to several Italian farmers who have in the 1990s continued to tend a thriving market garden operation. The produce, sold to the surrounding community, has established a direct relationship between producer and consumer in the middle of a highly developed urban area.

The stimulus to turn formerly vacant land to productive use of this kind lies in provincial land tax laws. Under the Ontario Assessment Act, land that is in use for agriculture is exempted from full land taxes levied by the local municipality. This act was introduced by the province to protect farmland from the sky-rocketing taxation that always follows urban development. On this basis the oil companies in the 1980s were paying Canadian $200 per hectare for the unbuilt land used for farming rather than the industrial rate of $980 per hectare.[59] The municipality, furious over lost tax assessment, took them to court, but was unsuccessful in its efforts to force them to pay full industrial rates, and the market garden has continued to flourish.

The anomalies of this case raise some interesting issues. A tax loophole provided an opportunity for turning unused industrial land into a productive, visually pleasing, self-sustaining landscape. It upgrades the general quality of the urban environment and has had direct benefits to the surrounding community in 'pick-it-yourself' fresh produce. Connections between agriculture and the city establish potential new patterns of landscape development in urban areas. At the same time, this inadvertent phenomenon may be seen as an unrealistic mechanism for achieving these ends in the existing political framework, where a municipality is denied the taxes it needs to provide public services. There is a further anomaly in the fact that if the oil companies were required to pay full taxes, the primary stimulus for farming their vacant lands would disappear. In this case, the land might either revert to the status of urban blight or be subjected to 'landscaping' to improve visual amenity.

This example points to the fact that intelligent and creative policies are needed to create the useful and productive landscape that will help provide cheap sources of food

Figure 5.5 *The need to protect productive soils and natural features and processes as cities grow involves issues of energy and compact urban form, ecology, soil productivity and rural traditions, and finding new ways of establishing a new kind of urban that is an integration of city and countryside*

Source: *Royal Commission on the Future of the Toronto Waterfront*, Regeneration, Toronto: *Minister of Supply and Services*, 1992

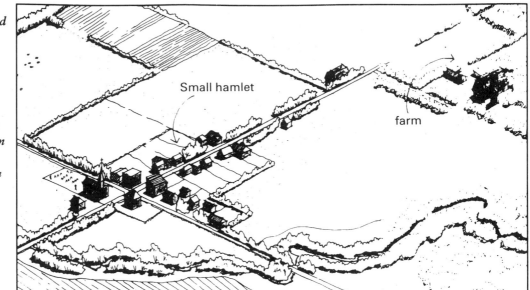

a   *The traditional rural village and surrounding countryside*

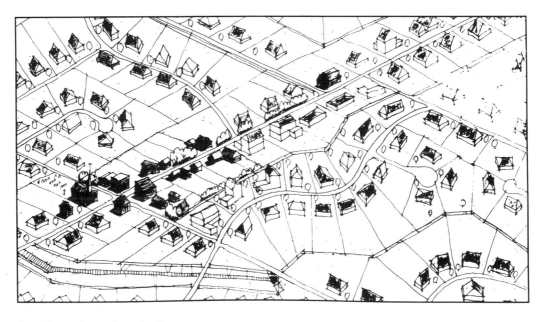

b   *The traditional rural village and countryside typically get swallowed up by conventional subdivisions*

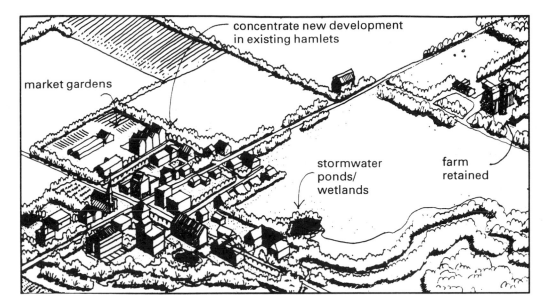

c *Clustering new development around the village leaves many rural functions, streams and woodlands intact*

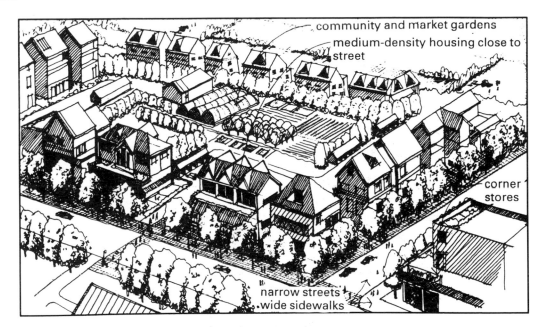

d *Mixed use development can be associated with small-scale market gardens and other rural-based occupations such as plant nurseries, pottery and crafts and recycling operations* and point the way to an alternative approach to urban design. For instance, there is a need to protect productive soils and rural traditions on the urban fringe in new development, thereby establishing a new kind of 'urban' that has an investment in the land, rather than traditional development that first confronts, then edges out, land-based industries. It is, in effect, a conflict of two incompatible cultures where the one inevitably degrades the other. Most of the arguments that are given for developing agricultural land are based on the need of today's farmers to sell land in order to survive or retire. Thus, productive land is relegated to a real estate trading commodity. Alternative strategies will depend less on the protection of large-scale farming operations that properly belong beyond city limits, and more on the marrying of development and small-scale and complex economic initiatives. These might include market gardens, greenhouse permaculture, allotments, small mixed farm livestock operations, plant nurseries, pottery and crafts, recycling operations and reforestation for servicing wood-based industries. The need for alternative planning strategies that reflect the economic, social and environmental needs of this changing countryside has implications for the developing as well as industrial nations. A 1985 study of food production in urban Kenya found that urban farming is of widespread economic importance to the survival of the poor. Although high- and middle-income people can benefit from low density residential planning, which enables them to practise backyard farming legally, low-income urban neighbourhoods are zoned for high density housing, which denies urban farming for those that need it most. The study suggests that alternative sanitary technologies, based on low water consumption, could be used to develop alternative housing layouts that incorporate farming and livestock.[60]

This multi-purpose and integrated planning approach may be found in the urban forest parks of Zurich, Switzerland, described in Chapter 3 (pp. 126–8). Here we find small-scale commercial agriculture being practised within the city's parks system. Farmers rent space on the common lands surrounding the city and use the land for crops, pig farming and related agricultural pursuits that can be carried out on a small scale. The recreational trail system of the park provides access through these areas as well as through the managed forests. In both Dutch and Swiss parks the concept of farming on a residential and commercial scale has been integrated as a basic function of the parks system. On both these scales farming is privately undertaken with the city providing and leasing out the land. The city's parks, therefore, function as food producers and for leisure and are administered on this basis. The productive processes of food production, of well-kept gardens and soil management become another visible element of urban processes. They enrich the urban experience and provide the basis for an aesthetic that comes out of a true application of sustainable principles linked to the land.

Neighbourhood parks and urban spaces based on community need can be used for community gardens, an activity that has the advantage of being under local control and

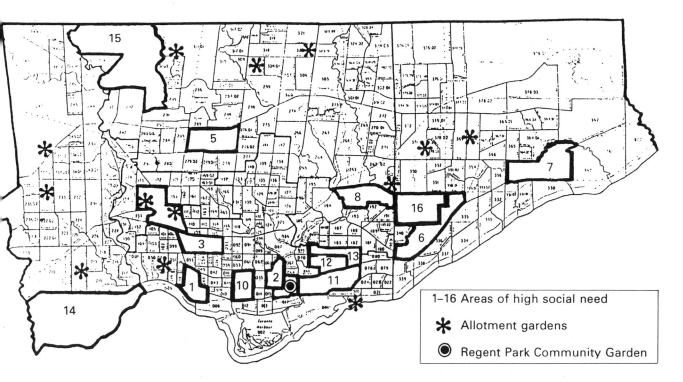

Figure 5.6 *Map showing areas of high social need and officially designated allotment gardens in Metro Toronto*

Note that the opportunities rarely coincide with the areas of greatest need. The municipal parks are not designated as food-growing areas

Source: *Social Planning Council of Metropolitan Toronto, 1985, cited in Michael Hough and Suzanne Barrett, People and City Landscapes, Toronto: Conservation Council of Ontario, 1987*

supervision. Livestock such as sheep, feeding on the city's grasslands while maintaining and fertilizing them, can provide the basis for a low labour-intensive and balanced urban environment. This, as we have seen, existed in the past and can still be found where necessity or long-standing tradition dictates.

These examples show what long tradition and intelligent planning at the municipal level can achieve – a multi-functional self-sustaining landscape that provides social, environmental and economic benefits. A key concept here is the idea that parks systems can be, at least in part, economically sustaining, contributing in ways other than recreation to the public good. In the context of the realities of poverty and hunger in cities around the world, there are potential social costs of not doing so. However, the ever-expanding food banks that have come to reflect one of the major issues of contemporary western society are only a temporary solution and may, in fact, exacerbate the problem by providing no long-term solutions to the major social issues of dependency, lowered morale and self-esteem. An examination of poverty and food-growing opportunities in Metropolitan Toronto compared the areas of high social need in the city with the location of city-run allotment gardens. Figure 5.6 illustrates that the opportunities to grow food rarely coincide with areas of greatest need.[61] Yet there is often general opposition to the notion that parks can and should, serve such functions. As we shall see on pp. 238–43, these issues have given rise to many community-based initiatives that include the International Federation of City Farms.

Food plants and design quality

The urban or suburban farmer who is producing food as a part of his livelihood is enriching and maintaining soil fertility, and making land productive with considerable public benefit in produce and amenity value. It is the kind of pleasing environmental quality that we have long accepted in the working rural landscape as a matter of course. But the concept of productive urban land also brings to bear the notion that plants that can be eaten can also have a role in landscape design. The observer of most city gardens cannot fail to notice that few if any have the slightest nutritional value. Conventional landscape planting has no place for cabbages, runner beans, squash or any other edible plant. But these have aesthetic qualities, too, in texture, form and colour if we stop to look at them with a designer's eye. The consciously designed landscape of the city, as we have seen earlier, is one that has been created by a leisured class that has no desire or need to grow its own food. Food plants, in fact, are usually relegated to the allotment garden. The ornamental flowering crab, cherry and almond replace the plants that used to produce real fruit, ensuring that their purpose in the accepted design idiom remains solely an aesthetic one.

However, the opportunities for using edible plants are just as great as those that are purely ornamental. Examples of the hundreds of trees, shrubs and groundcover plants that produce edible produce for the residential gardener have been well documented in Rosalind Creasy's book, *The Complete Book of Edible Landscaping*.[62] The potential and implications for design, however, go beyond the private garden and into the larger public domain of the city itself. Tree planting along city streets can include fruit-bearing species. Orchards can be grown on unproductive land and provide a return in city-grown fruit. The city of Adelaide, Australia, grows olive trees which are harvested by people who use the olive oil and preserve the fruit.[63] Village Homes, an innovative, energy conserving development in Davis, California, grows fruit trees along its streets and bikeways, including orange, crab-apple and nut trees. In addition, apricot and almond orchards and several vineyards grow in the community's public spaces. A substantial percentage of the food consumed by Village Homes residents is produced literally in their own back-yards.[64] In the Netherlands old orchards that have been encroached on by urban development are integrated into recreation areas with picnic areas and children's playgrounds. Vines grown on trellises for shade in public places, on buildings, terraces and in gardens can be selected to produce crops such as grapes, beans and scarlet runners that have colourful flowers and strong visual form.

The aesthetic implications for design also have to do with the very nature of multi-cultural society discussed earlier in this chapter: 'Multi-culturalism is not about what landscapes should look like. It's about process and an understanding that no one aesthetic is better than another.'[65] The cultural differences between people of different races is another factor that militates against commonly accepted notions of what is aesthetically appropriate and what is not. Expressions of beauty or symbolic important to the Chinese community may be regarded as aesthetic kitsch by the designer grounded in the conventions of the European design tradition. The potential imposition of such aesthetic values on multi-cultural urban communities is a crucial factor in determining the role of the designer in the modern city. It requires, therefore, very different values that have to do, first, with an awareness of the different ways in which people see the world and, second, with a recognition that different cultural groups are the best judges of what is appropriate for them and what is not. What is needed, in effect, is a response to the city's environment that supports its inherent diversity, rather than negating it with outdated aesthetic doctrine; one that provides a framework where different human needs can be expressed in their own ways.

Community action and urban form

The commonly held belief that parks should be provided at public expense by authorities and with little direct public involvement is a legacy of the past that is being re-examined in many cities. This belief dates from the growth of the industrial city in the nineteenth century and the development of the public park by men of social conscience who saw parks as an essential component of urban reform. In England, J.C. Louden's writings advocating public parks contributed to the support by the middle classes of the concept of gardens for the less fortunate. Andrew Jackson Downing and Frederick Law Olmsted in the United States both saw contact with nature as a source of pleasure and benefit to society and a necessary way to improve the cities.[66] Olmsted himself was convinced that:

> the larger share of the immunity from the visits of the plague and other forms of pestilence, and from sweeping fires, and the larger part of the improved general

health and increasing length of life which civilized towns have lately enjoyed is due . . . to the gradual adoption of a custom of laying them out with much larger spaces open to the sunlight and fresh air.[67]

Today, we recognize that the nineteenth-century Romantic view of the park as a piece of natural scenery, separated from the city, for contemplation and spiritual renewal no longer has the validity it may have once had. The whole physical, technological and social structure of urban life has changed. As I discussed in Chapter 3 (pp. 137–63), changing economic conditions, new demands and attitudes towards recreation and reduced park priorities have made it increasingly difficult to maintain the existing parks systems acquired over more than a century, or to expand them. There is, in addition, a radical shift in the way people view authority and government. The late 1980s and 1990s have seen a dramatic and worldwide growth of public interest and concern for the environment. A shift in the balance of power in favour of greater public participation in the affairs of the city is occurring – one that is based on positive action, rather than on the negative reaction that was typical of the 1960s and 1970s. The days when expressways, such as San Francisco's Embarkadero and Toronto's Allen Expressway, were being cut through city neighbourhoods and creating havoc with its fabric and environment have long gone. The demand for greater control over one's life and surroundings on the part of urban people is reflected in a changing view of parks and the city environment as a whole. New Yorkers feel personally possessive and protective about Central Park. Olmsted himself referred to it as a people's park, his original purpose being to create a public place rather than an aristocratic reserve. This attitude during the 1980s gave rise to a highly vocal public debate with respect to its restoration. As we saw in Chapter 2 (pp. 51–70), in cities around the world concerned citizens have initiated tree-planting projects in valleys and degraded lands, and the ecological restoration of urban rivers and their watersheds has brought a new era of co-operation between the public and government in achieving long-term environmental and social goals.

The city farms

The determination among people at the grass roots to be participants and initiators in decision-making where it affects them and their neighbourhoods, and the increasing incapacity on the part of public authorities to provide the amenities that have traditionally been their responsibility, are creating new conditions in cities. Experience in Britain has shown that the physical decay and social needs prevalent in cities are best tackled by those who have their roots, their families and their futures in the neighbourhood. The development of an alternative type of park, the city farm, was begun

Plate 5.6 *The City Farm, Kentish Town, London*

A new kind of park created through community action, where local children can learn about horses through direct experience in the local park. In addition, many barnyard animals, including chickens, goats and sheep are looked after by the children and adults

(Photo: Inter-Action Advisory Service)

Plate 5.7 *Part of the Kentish Town city farm is devoted to allotment gardens looked after by the elderly members of the community*

Leased from British Rail, the site was originally a derelict piece of railway land which has now become a productive multi-functional park through the efforts of the community. It is, in addition, economically self-sustaining, making no demands on the public purse

(Photo: Inter-Action Advisory Service)

in the early 1970s. At that time it aimed to bring derelict land back into use for the benefit of local communities. The city farm used farm animals as a central part of their activities that also included crafts, theatre, auto repair workshops and other activities. The Kentish Town City Farm was one of the first to be launched in 1972, and in 1976 the City Farm Advisory Service was set up with government funding. From then on the movement grew steadily, and by the 1990s there were some sixty city farms in urban areas across the United Kingdom.

The successor to the City Farm Advisory Service, the National Federation of City Farms, acts as a support and development organization. Its tasks are to facilitate planning approvals, help obtain funding, establish links with farmers and provide expert help and advice to communities. Its major aims are to create opportunities for people to enrich and develop their own lives by active participation, provide employment and work experience and make a positive contribution to conservation of the land by encouraging organic farming, gardening and forestry. The Federation receives a Home Office grant, through the Voluntary Services Unit. In addition, its income comes from membership fees, earned income including consultancy, conferences and sales of publications, charitable trusts and donations and sponsorship from private companies.[68] The city farms themselves, apart from occasional funding from the National Federation are self-sufficient. Their income derives from a wide variety of sources, which include local authorities, companies, charitable trusts and donations and earnings from the farms such as riding school fees and eggs. With respect to employment, the city farms have created over 300 jobs. Much has also been done to create training opportunities for people with mental and physical disabilities in developing the farms as successful businesses. Thus, a significant contribution is made to the resources of the various communities.[69]

Contrary to the petting zoos, popular in many cities, which send out the wrong signals about the nature of farm animals and farming generally, there is a basic emphasis on active involvement rather than on observation. There is also a great diversity of activities. Many city farms, for instance, keep a variety of farm animals including sheep, horses, cows, goats, chickens and rabbits. There are garden plots with individuals sharing facilities and communal land management. Many include shops, craft workshops and riding centres, or retain outlets for horticultural produce. Some of the larger and more established farms now have fully-fledged community businesses that generate sufficient income to cover extra staffing costs. Sanitation and health requirements are monitored by health inspectors who make regular visits to the community.

The city farms concept, therefore, has provided a basis for community revival in depressed urban areas and an educational link with rural occupations by enabling practical farming demonstrations to take place within urban neighbourhoods. As an alternative framework for urban parks, the concept is significant. It returns derelict land to productive use; and through community effort it is self-supporting, offering facilities

Table 5.4 *Comparative costs of local authority park and urban farm*

	Statutory authority	Voluntary agency
Name	Lisson Green Estate Playground	City Farm 1
Date of completion	31 July 1972	1 July 1974 but continuing development
Capital cost	£75,000 plus demolition costs	£5,690
Vandalism cost	Approx. £24,000. Building entirely vandalized	Approx. £20, 4 break-ins
Closure period for refitting	8 months	None
Running costs 1st year	£10,176 (direct fees for equivalent of four workers exclusive of on-costs, back-up staff, security and maintenance)	£4,200 (inclusive of on-costs, back-up staff and maintenance)
Source of finance	Ratepayers	Earnings from users plus donations
Construction agency	Westminster City Council, Eng. Dept	NUBS
Location position	Border of council housing estate	Border of council housing estate
Size	Approx. 1.2 hectares	Approx. 1.2 hectares
Child pop. in area	2,500	2,500
Facilities	Play hut, football pitch, kick about area, 3 mounds with swings, slides and nursery area	Stables (11 horses), tack room, indoor riding school, pony club, community gardening for elderly, greenhouse, 20 kitchen gardens, caretaker's house, storage block, household repair workshop, auto-repair workshop, farmyard (58 animals), tea room, meeting place/rehearsal room
Adult visits p.a.	Nil	19,000 estimate
Child visits p.a.	35,000	49,640 organized plus 25,000 random
Mode of construction	Comprehensive redevelopment; taxpayers' investment	Renovation with voluntary labour, donations in kind from local industry
Mode of organization	Council Eng. Dept (until May 1974, after the vandalism, handed over to voluntary organization with council funding)	Voluntary organization with users' management committee for each activity

Source: 'City Farms', Inter-Action Advisory Service pamphlet, London, n.d.

otherwise unavailable in inner-city areas. It adds nothing, therefore, to the burden of public expenditure for city parks, which in many housing developments are simply not being provided. In contrast to the normal city park, running costs are considerably less. Vandalism is reduced to a minimum as a consequence of community action and the unattractive targets of rough-hewn sheds and paddocks built by local labour. An interesting comparison was made in the late 1970s between the capital and running costs of a community urban farm and an adjacent London local authority park (Table 5.4). While this comparative analysis has not been revised since that time, it still provides an indication of the value that this form of public space provides. The lessons are self-evident. The comparison shows that the community farm is far less costly to build and maintain than the public park under the city authority. It is socially more viable, educationally more pertinent and physically more diverse. This example of a city farm in operation also demonstrates that urban wasteland is a resource out of place and that its creative and productive use does not necessarily depend on high capital investment.

The implications for urban design and management are clear. A new approach to our concept of the urban landscape is needed, requiring radical conservation measures to ensure the future environmental and social viability of cities. What we are concerned with is a design philosophy that integrates the ideals of urbanism with nature and rural skills and values. It brings us closer to the land and the biological systems from which urban people have been alienated, and gives us the practical tools with which to sustain ourselves in the future. The principles of productivity and diversity, the integration of environment and cultural diversity in the planning, design and management of urban landscapes flow from this philosophy. What logically follows is a view of urban land as functionally necessary to the biological health and quality of life in the city. Outmoded standards and design criteria for parks must be revised to permit a wider view of their purpose and function in the twenty-first century. We need a policy for urban land as a whole that encourages the creation of both commercial and community gardens, makes productive use of currently wasted energy and land resources, encourages the perpetuation of self-sustaining urban spaces, provides real economic benefits to the needy in times of economic depression and high unemployment, and contributes to the maintenance of cohesive and stable neighbourhoods. As working environments, parks must also be seen, to a greater or lesser extent, as economically self-sustaining, providing returns for investment in food and services and contributing to the evolving needs of people.

For many, urban farming is an alternative way of renewing contact with the land and nature through therapeutic and healthy work. For others it is an increasingly necessary method of obtaining food at a reasonable cost. This is particularly significant in developing countries where the imperatives of employment, food, recycling of every scrap of materials and economic survival in the burgeoning cities have forced immigrants on to government owned or waste land. At the same time it can be argued that problems of poverty and homelessness are common to all cities, irrespective of how 'developed' they have become. People everywhere should be familiar with the broad scientific principles of the production, handling and use of the food that appears on the table. It teaches through direct experience something of rural occupations and the basis for alternative sustainable values. The concept of productivity in urban design terms, therefore, has wide implications. Drawing its inspiration from ecologically and socially based management practice and understanding of human aspirations, it deals with the maintenance of diverse natural and cultural environments. It is from this that its aesthetic inspiration is derived. This inherent variety and richness of purpose is the imperative that can shape a new vernacular and provide the foundation for the alternative design language we seek.

CLIMATE: MAKING CONNECTIONS

•

In Burma once, while Bishop Prout
Was preaching on Predestination,
There came a sudden water spout
And drowned the congregation.
'O Heav'n' cried he, 'why can't you wait
Until they've handed round the plate!'[1]

INTRODUCTION

The notion that we are at the mercy of the weather is aptly illustrated in this childhood poem by Harry Graham. The interacting variable forces of wind, precipitation, temperature, humidity and solar radiation are the great climatic forces that have shaped the world's bio-regions, and to which, historically, all life forms including the human race have adapted. Indeed, it can be said that climate, more than any of the natural systems we have examined in this inquiry, transcends all the boundaries of nature and human activities. It pervades and influences water, plants, wildlife and agriculture. It is the fundamental force that shapes local and regional places and is responsible for the essential differences between them. At the same time human settlement has modified micro-climates to suit particular needs and local conditions. Human comfort, and in some cases survival, have depended on the skill with which building and place-making have been able to adapt to the climatic environment. The modern city has had a greater impact on this environment, on living conditions and attitudes, than at any other time. The old arts of creating felicitous outdoor places that take advantage of climatic elements and the material resources of the landscape seem to have been lost. As pressures for energy conservation and the need for civilizing places to live in become more urgent in the last decade of the twentieth century, we must look to environmentally sounder ways of manipulating the climate of cities than the present total reliance on technological systems. My purpose in this chapter is threefold: first, to review the nature of urban climate, and to explore how the outdoors can usefully contribute to urban liveability and conserve the city's energy; second, to bring the various components of natural and human systems into an overall framework for design – seeing the pieces of the puzzle as a whole picture; and, third, to show how the principle of connectedness, embodied in the influences of local climate, has global implications at every level.

NATURAL ELEMENTS AND CLIMATE

The basic elements of climate – solar radiation, wind precipitation, temperature, humidity – are affected and moderated by the elements of the land, including topography and landform, water and plants. At a macro-scale, landforms create barriers to the movement of air masses. They affect moisture conditions on the windward and leeward sides of hills and mountains. They affect temperatures at different heights of land – temperatures decreasing with altitude. Landforms control the flow and temperature range of air by forming impediments and channels to movement. They create katabatic valley winds that move up during the day and flow down at night, settling in valley bottoms as pools of cold air. South-facing slopes concentrate solar energy and produce different micro-environments from shaded slopes, which affects the growth and patterns of vegetation.

Vegetation controls direct solar radiation to the ground and hence the heat radiated back from ground surfaces. A forest may absorb up to 90 per cent of light falling on it and in general reduces maximum temperature variations throughout the year. It may reduce wind speeds to less than 10 per cent of unobstructed wind and maintain more equitable day and night temperatures than non-forested land. It regulates the amount and intensity of rain reaching the forest floor and affects the deposition of rain or snow and humidity. It reduces glare from reflective surfaces since leaves have a low reflective index.

Water has a profound impact on climate control. Large bodies of water absorb and store a high percentage of solar energy. They heat up and cool much more slowly than land masses and so act as moderators of temperature on land through the ventilation of onshore breezes. The process of evaporation of water converts energy from the sun into latent heat, reducing air temperatures and acting as a natural air-conditioner.

URBAN INFLUENCES ON CLIMATE

It is predicted that the energy and resource requirements of cities will, in the foreseeable future, affect not only local but regional and macro-climates.[2] Global warming, acid rain and other now familiar and much discussed issues all originate in cities. A great deal can be done to influence urban climatic conditions locally, however. To do so, we need to examine and understand the influences that affect urban climate.

It is quite apparent that the climate of cities is markedly different from rural areas. Various climatological studies have accounted for five major influences that affect urban climate, based on the fact that energy is the basis for the climatic differences between city and countryside.

- *The difference in materials in urban and non-urban environments.* The impervious surfaces of city streets and paved spaces and the stone and concrete of building surfaces, store and conduct heat much faster than soil or vegetated surfaces. In addition, urban structures are multi-faceted. Roofs, walls and streets act as multiple reflectors, absorbing heat energy and reflecting it back to other surfaces, so the entire city accepts and stores heat. It becomes, therefore, a highly efficient system for heating large quantities of air throughout its volume. In the countryside, on the other hand, heat is stored mostly in upper layers. In a wooded area the canopy receives and retains most of the heat, while lower levels remain relatively cool. City temperatures are generally warmer than the areas outside. Chandler has found over thirty years that the average temperature in London was several degrees higher than in outlying areas.[3] An illustration of the considerable contrasts in temperature regime of various materials is given by Miess. In relation to a given quantity of energy received, open water is the most constant. Between early morning and midday its increase in temperature may be no more than 3 to 4°C. By contrast, during the same period asphalt may have a temperature increase of 30°C. The temperature of grass may increase by 20°C, but at the same time its temperature drops to much lower levels at night.[4]
- *The much greater aerodynamic roughness of built-up areas than in the countryside.* The arrangement of tower blocks placed individually on their own sites presents a much rougher surface than the open country. This has the effect of slowing down prevailing winds and increasing localized gusts at street corners and around tall buildings, and diminishing the cooling power of wind in summer.
- *The prodigious amount of heat energy pumped into the city atmosphere from heating and cooling systems, factories and vehicles.* Central air-conditioning for residential buildings grew from 9 per cent of US households in 1969 to 15 per cent in 1973,[5] and the numbers have continued to grow. In winter large amounts of heat are lost to the exterior and in summer, air-conditioners cooling interior space pump hot air to the exterior, making the problem of high temperatures worse.
- *Problems resulting from precipitation.* Rain is quickly carried away by storm sewers and, in northern climates, snow is usually cleared from city streets and pedestrian areas. Evaporation converts radiated energy into latent heat, which acts as a cooling process. In the countryside, moisture either remains on the surface or immediately below it. It is thus available for evaporation and cooling. But in the city, the absence

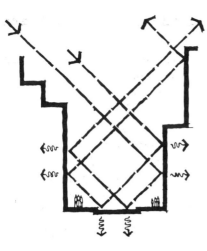

Figure 6.1 *In the city, vertical walls reflect solar radiation to the floor and walls of buildings. Impervious surfaces in walls and floors accept and store heat*

Source: *William R. Lowry, 'The Climate of Cities',* Scientific American, *August 1967*

Figure 6.2 *In the countryside, solar radiation is reflected back to the sky due to lack of vertical impervious surfaces*

Tree canopy retains heat, while lower levels remain cool

Source: *William R. Lowry, 'The Climate of Cities',* Scientific American, *August 1967*

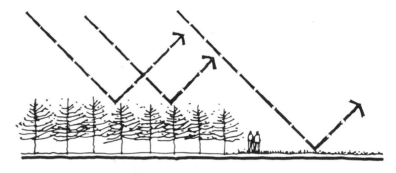

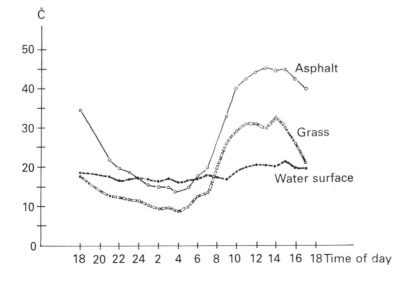

Figure 6.3 *Surface temperature of materials*

The different physical characteristics of the various surfaces exposed to radiation give rise to a very contrasting temperature regime

Source: *Michael Meiss, 'The Climate of Cities', in Ian C. Laurie (ed.),* Nature in Cities, *New York: John Wiley, 1979*

of moisture inhibits evaporation. And so the energy that would have gone into the process of cooling the environment is available for heating, a decisive factor in energy exchange.[6] The capacity of building materials to store heat is greater than that of air by about 1,000 times, and the process of transfer by means of air particles into the atmosphere is much less efficient than evaporation. Only over open water and areas of vegetation does the process of evaporation become fully effective.[7]

- *Air quality*. It is estimated that the major air quality issues of the 1990s are likely to be ozone, particulates and atmospheric carbon dioxide.[8] Increased atmospheric carbon dioxide likely will lead to increased air temperatures and exacerbate ozone problems, with the major emission sources of all three being automobile and industrial processes. Ozone is formed by a photo-chemical reaction of nitrogen oxides and volatile organic compounds in ultraviolet sunlight and moisture, and affects the respiratory tissues and functions in humans. A heavy load of solid particles, gases and liquid contaminants is carried in the urban atmosphere. There are ten times more particulates in city air than in the countryside, which reflect back incoming sunlight and heat, but also retard the outflow of heat.[9] A high volume of particles in the atmosphere reduces the penetration of short-wave radiation in the ultraviolet range. This is biologically important to the production of certain vitamins and the maintenance of health. Motor vehicles are the major source of carbon monoxide in North America and can reduce the oxygen carrying capacity of blood. The higher the atmospheric concentrations of carbon monoxide the more serious the health effects.[10] Dilute sulphuric acid emissions are primarily caused by fuel combustion from stationary sources including coal. It is an irritant of the respiratory system which, when breathed in, causes bronchial constriction resulting in increased respiratory and heart rate. The consequences of air pollution on health have been found, in effect, to be serious, particularly in developing countries where air quality standards may be minimal, or non-existent. A recent study in Poland, for instance, has found new evidence linking air pollution with widespread genetic damage and birth defects. A group of organic compounds in smog caused by burning coal (a major source of energy and domestic heating in the region) called polycyclic aromatic hydrocarbons (PAH) exert a damaging effect on the genetic material in the cells of people breathing air, drinking water and eating food. A comparison between a satellite image of pollution dust in Upper Silesia and a map showing the frequency of cancer in women shows a nearly identical correlation between the two.[11]

The urban heat island

Buildings, paving, vegetation and other physical elements of the city are the active thermal connections between the atmosphere and land surfaces. Their composition and structure within the urban canopy layer, that extends from the ground to above roof level, largely determine the thermal behaviour of different parts of the city.[12] Warmer air temperatures in cities compared to surrounding rural areas are the primary characteristic of the urban heat island. This phenomenon has been studied in some detail by

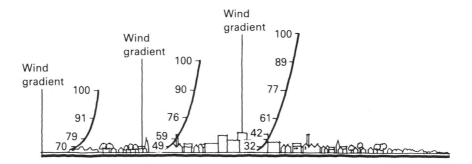

Figure 6.4 *Typical wind profiles over built-up area, urban fringe and open sea*

Increased aerodynamic roughness of built-up areas causes rapid deceleration of wind compared with open countryside. It has been calculated that wind velocity within a town is half of what it is over open water. At the town edge it is reduced by a third

Source: *Michael Meiss, 'The Climate of Cities', in Ian C. Laurie (ed.), Nature of Cities, New York: John Wiley, 1979*

Figure 6.5 *The urban heat island*

Smog dome over large cities occurs periodically due to urban activities. Air rises over the warmer city centre and settles over cooler environs so that a circulatory system develops. The dome and its effect on city climate may persist until wind or rain disperses it

Source: *William P. Lowry, 'The Climate of Cities', Scientific American, August 1967*

Figure 6.6 *Generalized cross-section of a typical urban heat island*

Source: *T.R. Oke*, Boundary Layer Climates, *New York: Methuen, 1987*

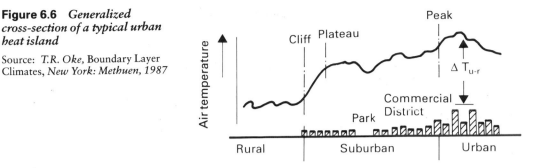

climatologists and is the result of the complex effects of the city's processes on its own climate. Lowry describes it as follows.[13]

Assuming a large city set in flat countryside with no large bodies of water nearby, the rising morning sun strikes the walls of its buildings, causing them to absorb heat. In the countryside, however, the sun's radiation is largely reflected off the surface with little heat absorbed. As the morning advances, the countryside begins to warm up, but the city already has a large lead towards maximum temperatures. The warm air in the city centre begins to rise and gradually a slow air circulation is established with air moving in, rising in the centre, flowing outwards at high altitudes and settling again in the open countryside as it cools. Near midday, temperatures inside and outside the city tend to equalize so that the cycle is weakened. As the afternoon passes, and the sun sinks, much of its radiation is reflected off the countryside but continues to strike building walls directly. Thus the circulation of air is repeated.

During the night the roofs and streets and other hard surfaces of the city begin to radiate heat stored during the day. A cool air layer is likely to be formed at the rooftop level. A stratification of air develops, inhibiting warmer air between buildings from moving upwards. The rural areas, however, cool rapidly at night, due to light winds and unobstructed radiation to the night sky. Although both city and countryside continue to cool during the night, by dawn the city is likely to be 4 to 5°C warmer.

The following day the heat, smoke and gases from the city are contributed to the heat being generated by radiation. The rising air also carries with it suspended particles of dust and smoke. Over time a dome-shaped layer of haze is formed over the city. At night the particles in the dome become nuclei on which moisture condenses as fog: this fog gets thicker by downward growth and eventually reaches the ground as smog. Smog inhibits cooling of the air and helps to perpetuate the dome by preventing particles from moving out of the system. In the absence of wind or heavy rain, the smog continues to build up. Since less sunshine can penetrate to warm the city in winter, increased fuel consumption adds to the smog build-up.

The process, in effect, is self-perpetuating and is responsible for the severe climatic problems that many cities face. For instance, approximately 3 to 8 per cent of the current demand for electricity for air-conditioning in the United States is used to compensate for the heat island effect because city temperatures have increased by about 1 to 2°C since 1950.[14] The effects of urban heat islands also have wider implications since urban temperatures are increasing worldwide. Comparisons of temperature data from paired urban and rural weather stations suggest that the recent warming trends are due to the heat island effect rather than changes in regional weather.[15]

PROBLEMS AND PERCEPTIONS

Mechanical climate control

The search for optimum human climates has been a continual process, particularly in those mid-latitude climates that are less extreme or predictable. Over the last two centuries remarkable changes have been made in the living environment through improvements to mechanical equipment. James Burke describes the chain of events that led to the invention of mechanical air-conditioning, by Dr John Gorrie in 1850.[16] As a physician working in Apalachicola, a small Florida cotton port situated on the Gulf of Mexico, Gorrie had been asked to report on the effects of climate on the population with a view to a possible expansion of the town. Among his recommendations was the need to establish a hospital to treat the fever that sailors and waterside workers endured every summer – an illness that was endemic to the town. Gorrie had noticed that malaria seemed to be connected with hot, humid weather. He began to solve the problem by using ice, circulating the cool air round the hospital wards by means of fans. Since ice was prohibitively expensive, he resorted to a known method for absorbing heat from surrounding gases. He constructed a steam engine to compress air, which, when cooled, rapidly expanded, and could then be circulated round a room. Subsequently, ice-making and refrigeration machines evolved from this invention and were used to transport food in ships from Australia and also for making German beer. This was followed by the domestic thermos flask and refrigerator for keeping food and drink cold. These inventions became the precursors for the modern air-conditioning unit. One of the first large installations for comfort control of office space was the 300 tonne unit installed in the New York Stock Exchange in 1904.[17] The mechanical climate control of buildings has had a number of fundamental effects on the modern city.

- It has freed buildings from the constraints of weather that were originally imposed upon it. Stylistically, modern architectural form has become an event in its own right, its design responding to the constraints of mechanical engineering rather than to the constraints of site and climate. Modern air-conditioning has permitted the development of the megastructure: great interconnected interior complexes, whose heating, cooling, humidity and daylight are entirely dependent on mechanical systems.

- It has contributed to the radical changes in urban form that have taken place since fossil fuels and other forms of energy have become abundant. The city turns its back on an outdoor environment that has become increasingly unlivable; an environment polluted by dust, smog and exhaust fumes, and alternatively swept by winter winds and cooked by summer heat.

- The preoccupation with internal climate has the effect of denying a climatic role for exterior space. Air-conditioning screens out the products of industrial processes – the chemical pollutants and dust that threaten public health. Unhealthy outdoor climates generate greater reliance on safe, controlled interior ones, and so more and more development provides interior space for urban activities. The subterranean shopping mall and links below the surface of the city is the modern alternative to the open air market.

- Its effects on lifestyles and perceptions of the environment have been profound. Urban life has become a series of air-conditioned experiences. The home, the office, the school, the bus that takes the children there, the movie theatre, have all been sealed off from the outdoors. It creates a world of its own; separated from the increasing problems of health and comfort in the world outside. It is remarkable how much energy and effort is expended to provide climatic comfort indoors, while at the same time maintaining such unrewarding environments outdoors. At one time it was even proposed that whole cities should be covered by geodesic domes: a suggestion that was seriously discussed for some cities, and which takes the problem a step further to the ultimate technological solution. One of the reasons for the lemming-like flight from the city on summer weekends is to escape from an oppressive urban climate and the air-conditioning unit for the clean air, breezes and sunshine at the summer cottage.

As I pointed out in Chapter 1, planning and design doctrine has traditionally been more concerned with conceptual ideologies of built form than with the determinants of natural process. Many early North American towns and institutions, following established planning ideologies, paid little heed to the extremes of climate in the regions in which they were located.

The effort to become independent of the variables of the environment have, by today's standards, been successful. Totally inhospitable places, from the Arctic to the Equator, can now become habitable, with mechanical heating and cooling systems providing uniform interior temperatures. But it has become abundantly clear that the present costs in energy to achieve such a goal are wasteful and, to a great extent,

a

b

Plate 6.1 *The preoccupation with artificially controlled climate within buildings: making pleasant outdoor environments has consequently been ignored*

(Photo: Steve Frost)

unnecessary. The air-conditioner provides evidence of this fact. The process of keeping cool inside in summer increases the already high temperature outside – a non-productive transfer of heat from one place to another. There are cheaper and more effective ways of achieving similar results. Since the outdoors comprises a large part of the city environment, it can contribute to the modification of climate. At the same time it will be apparent that some problems cannot be solved in the context of natural process alone. Air pollution, for instance, is a macro-problem requiring solutions at the source – the industries and vehicles that create it. This involves many technical and institutional issues that cross regional and national boundaries, and are beyond the scope of this inquiry. But there are positive aspects to what may be seen as an environmental problem when urbanism and nature are seen as a whole. Design, inspired by ecology and laced with a good dose of common sense, provides solutions at less cost and effort. The vernacular forms of older towns and urban landscapes are examples of adaptation that provide some inspiration and guidance for application today.

ALTERNATIVE VALUES

Macro-climate, moderated by landform, vegetation and water has, in various ways, influenced the location and nature of human settlement and uses of the land. Carter has observed that the role of the environment is determined primarily by culture rather than

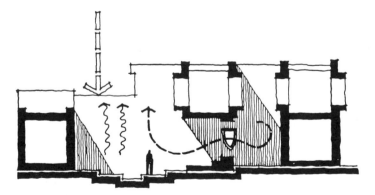

Figure 6.7 *The two courtyard house*

In this traditional Middle Eastern design, the deep, shaded courtyard is cool, the large courtyard is warm. The difference in air pressure induces a convection draught from the cool to the warm area. Water cooling jars placed in the passageway add to the cooling effect of the breeze

Source: *Allan Cain, Afshar Farroukh et al., 'Traditional Cooling Systems in the Third World'*, Ecologist, *vol. 6, no. 2, 1976*

the other way around.[18] There is no doubt, however, that humankind has responded in characteristic ways to climatic influences within its control. To be physically comfortable is a fundamental human need and a comfortable temperature lies in the range 20 to 24°C. People are affected by climate and react to it even though the response may not be conscious. There is a marked difference in the use of urban places at different seasons. On a winter's day people crowd the sunny side of the street. They seek out spaces protected from the wind; they gather where, on a hot summer's day, park seats and patches of lawn and trees provide shade or a cool breeze. The manipulation of natural and human-made elements of the environment and solar energy to create felicitous and healthy places to live and work in, has preoccupied urban people since the beginning of recorded history.

The business of keeping warm or cool in energy deficient societies is achieved by the necessity of accepting the limitations of the climatic environment and making the most of its opportunities. This has in the past, and still does in places where traditional technologies are maintained, been done with great sophistication and economy of means. The Maziara cooling jar, for instance, is a traditional water cooling and purification system, used in rural areas of Upper Egypt and other parts of the world for keeping liquids and perishable foods cool.[19] The action of evaporation absorbs considerable amounts of heat energy (580 calories of energy for every cubic centimetre of water evaporated).[20] Experiments have shown that with air temperatures ranging from 19 to

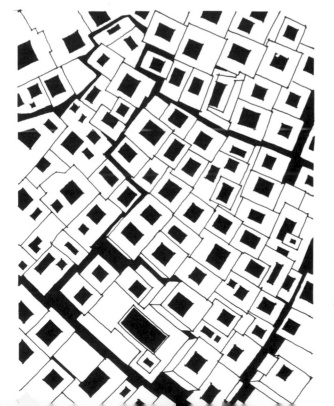

Figure 6.8 *Built form of towns in hot Mediterranean climates was often based on courtyard houses closely packed along narrow winding streets to maximize shade*

Plate 6.2   *A street in Istanbul*

Plate 6.3 *Built form in cool climates. Closely packed buildings and streets of the pre-industrial city, built to conserve energy, trap sunlight and exclude winds. Sedbergh, Cumbria*

36°C, water temperatures in the Maziara jar remains at a constant 20°C.[21] Cities built in hot dry climates took advantage of wind for ventilation and cooling. Rudofsky illustrates a dramatic example of natural air-conditioning in the Lower Sind district of west Pakistan, where specially constructed wind scoops, installed on roofs, channelled the prevailing wind into every room.[22] Some ingenious passive cooling systems in Iran, described by Bahadori, use wind towers which operate by changing air temperatures and thus its density. The difference in density creates a draught, pulling air either up or down through the tower and through the building.[23] Courtyard houses and the agglomeration of buildings along narrow streets, typical of Middle Eastern and Mediterranean towns, maintain coolness by trapping cool night-time air and retaining it by day. Many cities in Africa and Spain use awnings or arcades to shield streets from the midday sun. Cities along the Mediterranean coast of North Africa are sited so that their streets, laid out at right angles to the shore, funnel incoming sea breezes.[24]

Plants and water have long been associated with city courtyards and gardens to provide air-conditioning and places of delight. The Moorish gardens of the Alhambra are a particularly felicitous and well-known adaptation to hot dry conditions in southern

Spain, where the evaporation of water off tiled surfaces and the dappled shade of plants cool its arcaded courtyards. Buildings in cold climates have employed techniques to conserve heat. The traditional Inuit igloo, a perfect expression of adaptation, employs the highly insulating properties of snow and orientation of entrances away from wind to create a habitable micro-climate under the harshest conditions. The hemispherical shape provides maximum volume for minimum surface area, thus minimizing heat loss. It is said to maintain temperatures of 10°C when temperatures outside are −45°C.[25] The narrow winding streets, enclosed squares, closely packed buildings and courtyards of the pre-industrial city provide as good a demonstration as any of response to climate and energy conservation. The streets minimized the effects of winds and the open squares trapped sunlight.

In all successful climate control, the siting and organization of built elements and spaces, the use of landform, plants and water have achieved optimum environments for living. Faced with climatic extremes, traditional design methods greatly enhanced urban environments because there was no alternative. Adaptations of these technologies to the modern city are equally necessary if its climatic environment is to be improved. Given the limitations of industrial pollutants, the open places of the city have an important function to perform in the restoration of the energy balance. It is to the resources and the techniques available to achieve this that we must now turn.

SOME OPPORTUNITIES

Solar radiation and heat gain

The amount of incoming radiation into a city is dependent on its layout and the pattern of its buildings, streets and open air places. Where sunlight penetrates direct to the floor, for instance in places that have large plazas and wide streets, radiation is most effectively controlled by vegetation, in particular, trees. The capacity of the forest canopy to absorb large amounts of heat energy is considerable. Short-wave radiation in a closed canopy of maple can be reduced by 80 per cent on a clear midsummer day. The forest can also reduce maximum air temperatures by about 6°C below the temperature in the open.[26] In the city the greater the closure of a tree canopy the greater will be its air-conditioning effect on surface temperatures. Surveys carried out in Germany compared a well treed square with a comparable area without trees. The daily radiation balance in June of the treed area versus the treeless one showed a difference of 256 per cent.[27] Deciduous trees have the great advantage, in climatic regions that suffer from extremes of summer and winter temperatures, of providing shade in hot seasons and permitting the sun to penetrate to the floor in winter.

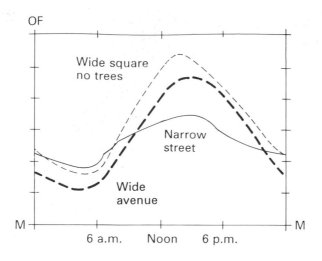

Figure 6.9 *Diurnal temperature variations in Vienna, 4–5 August 1931*

The graph shows the differences in temperature for a wide square with no trees, a wide avenue with trees and a narrow street

Source: *Gary O. Robinette*, Landscape Planning for Energy Conservation, *Reston, Va.: Environmental Design Press for the American Society of Landscape Architects Foundations, 1977*

Wall-climbing vines perform a similar function with respect to south-facing building wall surfaces. While attention may be focused on the considerable area of ground surface in cities requiring shading, it is easy to forget that vertical surfaces vastly increase the area subject to heat gain (see pp. 264–7). A German calculation indicates that there is an aggregate of some 50,000 hectares of vertical surfaces in German cities.[28] In energy terms the calculation suggests that vegetation on vertical surfaces can lower the summer temperatures of the street by as much as 5°C. Heat loss from buildings in winter can be reduced by as much as 30 per cent.[29] Biologically, the leaf is an efficient solar collector. During the summer the leaves are raised to take advantage of solar radiation, permitting air to circulate between the plant and the building. It cools, therefore, by means of a 'chimney effect', and through transpiration of the leaves. In winter, the overlapping leaves form an insulating layer of stationary air around the building. Even in climatic regions that are too cold for evergreens to grow successfully, summer cooling may still be an important factor, lending an energy-saving and biological validity to what is generally regarded as a decorative addition to architectural façades.

Rooftops in dense urban situations also receive a high level of solar radiation, paralleling the heat of ground surfaces. Rooftops constitute a large proportion of the city's upper-level environment in downtown areas and are, from taller structures, often highly visible parts of the urban landscape. Rooftop vegetation functions in the same way as it does at ground level. Rooftop gardens, therefore, can perform a functional role in climate control. The limitations to creating this kind of landscape relate to problems of structural support for soil and plants which, while they can be overcome in new projects, may constitute major problems for the large areas of existing roof areas in the city. There

Plate 6.4 *Plants covering building walls do more than look nice. In hot climates they can reduce city temperatures by as much as 5°C: a parking garage in Jakarta, Indonesia*

are also severe drainage, irrigation, nutrient and (in cold climates) frost problems to be overcome. Alternative strategies must, therefore, be found to combat one of the major climatic and visual problems of many urban areas. An examination of many old rooftops will reveal that fortuitous plant communities often gain a foothold. Mosses, grasses and, in places where a small amount of humus and water can collect over time, even adventitious shrubs and small trees colonize these unattended and forgotten places. The issues of fortuitous plant communities have been discussed in detail in Chapter 3 (pp. 101–6), but it is relevant here to pursue their role in climate amelioration. A research programme undertaken by the Parks Department in Berne, Switzerland, on a concrete garage roof, showed that certain plants can be grown on 7 cm of soil consisting of pea gravel and silt sand. Various sedum species were used and became well established after a year. Other experiments using grasses and climbers such as Virginia creeper (*Parthenocissus quinquefolia*) have shown that many vigorous plants can adapt to such environments with minimum soil depth or humus content.[30] The city of Dusseldorf, Germany, has regulations requiring large flat-roofed areas to have roof gardens. These and other examples in Europe suggest that hardy plants adapted to city conditions can survive in hostile environments at very little cost and without adding to the weight of existing or new building structures, and that the creation of a new and economical landscape at roof level can contribute to the climatic conditions of the city by reducing heat absorption.

Temperature controls

City temperatures are related to solar radiation and heat gain from urban materials. Temperatures are generally higher than in the countryside. This is accounted for by reduced evaporation, greater conductivity and heat storage capacity of building materials, variations in wind around buildings and the high proportion of airborne pollutants. Temperature can be controlled in several ways.

Water

One of the most effective ways of controlling local climate is through the evaporation of water into the air, particularly in dry climates. This is achieved in several ways: by direct evaporation from open water and by the evapotranspiration of plants. The high run-off coefficients of paved surfaces and the efficient removal of surface water by storm sewers, have effectively removed its availability for evaporation and cooling. This function is greatly assisted by reintroducing water into the city. It occurs by design where pools, fountains or artificial lakes are reintroduced into the city landscape. But it also occurs fortuitously after rainstorms where water forms ponds in parking lots, low-lying turfed areas and other 'poorly designed' places, where it becomes freely available for evaporation. Surface water introduced into the city's paved and unpaved places by impounding rainwater serves an important climatic function, as well as restoring hydrological balance, and providing recreation, wildlife and aesthetic enhancement. There are also pollution and erosion control advantages in doing so that have been discussed in more detail in Chapter 3 (pp. 126–8).

Plants

Plants evaporate water through the metabolic process of evapotranspiration. The cycle of water is carried from the soil through the plant and is evaporated from the leaves as a part of the process of photosynthesis. It has been estimated that on a single day in summer, 0.4 hectares of turf will lose about 10,800 litres of water by transpiration and evaporation.[31] The transpiration of water by plants helps to control and regulate humidity and temperature. A single large tree can transpire 450 litres of water a day. This is equivalent to 230,000 K calories of energy in evaporation which is rendered unavailable to heat surfaces or raise air temperatures.[32] Federer has compared the effectiveness of this evaporation by a tree to air-conditioning. The mechanical equivalent to the tree transpiring 450 litres a day is five average room air-conditioners, each at 2,500 K calories per hour, running 19 hours a day.[33] He also points out the important fact that the air-conditioner only shifts heat from indoors to outdoors and also uses electric power. The heat is, therefore, still available to increase air temperatures. But with the tree, transpiration renders it unavailable.

It will be obvious, therefore, that in energy terms a tree shading a house is more effective. It produces no unwanted waste products from the process of cooling, uses no electric power and continues to work better and better over the life of the tree. A numerical simulation of urban climate has suggested that where at least 20 per cent of an urban area in mid-latitudes is covered by plants, more incoming solar radiation is used to evaporate water than to warm the air.[34] These facts are borne out in Davis, California, where temperatures can reach 38°C. Various studies have shown that neighbourhoods with shaded narrow streets can be as much as 6°C cooler than those with unshaded streets. Moreover, a neighbourhood that is 6°C cooler uses only one-half the amount of electricity for air-conditioning than an unshaded neighbourhood.[35] Based on these findings, in 1977 Davis City Council unanimously passed an ordinance requiring a minimum of 10 per cent of a paved parking lot to be canopied by trees within fifteen years of the building permit's issuance.[36]

Climatic factors also determine management objectives. The ability of trees to act as sponges of carbon dioxide in the face of increasing concern and debate over global warming has greatly enhanced the attractiveness of urban tree planting. Analysis of the urban forest in Oakland, California, reveals it currently stores approximately 145,000 tonnes of carbon.[37] Future growth and planting of trees can add to that storage total if the amount of carbon sequestered due to growth and planting remains greater than the amount of carbon lost due to mortality. In addition, it has been calculated that large-scale planting and the use of light-coloured surfaces in cities have the potential to conserve about 2 per cent of the total production of carbon in the US.[38] The landscape value of trees in raising housing values or in making light industrial areas more saleable is also a significant factor.[39]

Wind

Of all the influences the city has on weather it is the presence or the absence of wind that has the greatest impact on the comfort of the local climate, as anyone who has walked the streets of a Canadian mid-western prairie city in winter will agree. There is less wind on average in cities than in the open countryside. Meiss states that within a town, wind velocity may be half of what it is over open water.[40] On the other hand, the existence of

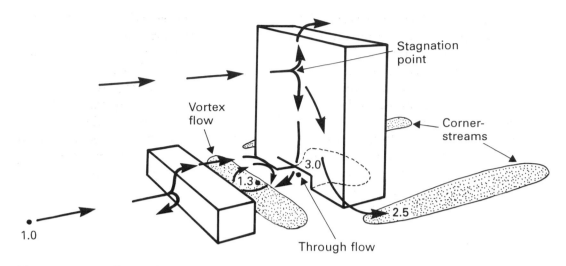

Figure 6.10 *Air flow in the vicinity of a tall building with smaller buildings upwind*

The stippled areas show areas of considerably increased wind speed at pedestrian level

Source: *T.R. Oke, 'The Significance of the Atmosphere in Planning Human Settlements'. In E.B. Wiken and G. Ironside (eds).* Ecological (Bio-physical) Land Classification in Urban Areas. *Ecological Land Classification Series, no. 3, Ottawa:, Environment Canada, 1977. Modified after Penwarden and Wise 1975*

free-standing towers separated by large open areas and the general layout of streets speed up winds locally, creating the unpleasant gusty windswept conditions and drifting snow which are typical of a winter's day. Wind affects temperatures, evaporation, the rate of moisture loss and transpiration from vegetation and drifting snow; all of which are particularly important to local micro-climatic conditions.

The impact of the wind environment around tall building complexes that have become typical of many suburban development projects is well known. The generally uniform low building layout of older towns, arranged along curved streets, provides shelter – the result of lower wind gradients at ground level. When winds meet a building that is considerably taller than its neighbours, the flow patterns changes. Air currents divide at about two-thirds or three-quarters of the building height, creating a down draught on the windward face, and highly turbulent conditions at ground level. Reduced air pressure on the lee side creates suction and high wind speeds around corners and through passageways under the building. Oke has found that since the force exerted by the wind increases as the square of its speed, a threefold increase of speed is associated with a ninefold increase in force; sufficient to knock down passing pedestrians.[41]

Many other studies on the effects of forest cover and shelter belts on winds have been made with respect to the speed of air movement, the protection afforded and the effect of wind barriers on heat loss from buildings. Tree stands are effective in slowing wind;

the greater the roughness of the ground surface the more wind velocities are reduced. The smaller an open forest clearing, the less turbulence at ground level there will be. Shelter belts may reduce winds by 50 per cent, the sharpest reduction in wind velocities extending ten to fifteen times the height of the trees on the lee side.[42] According to Olgay, a 32 k.p.h. wind can double the heat load of a house normally exposed to 8 k.p.h. winds.[43] The agricultural experiment station at Kansas State University has shown that the heating load on a house can be greatly reduced with the use of wind breaks.[44] Measurements of the effect of wind-break planting around unprotected residential buildings indicate that annual heating costs can be reduced by 10 to 30 per cent.[45] The Ontario Ministry of Housing has also calculated that landscaping around residential buildings in the temperate climate of southern Ontario is capable of producing energy savings in excess of 5 per cent.[46]

Air pollution control

As I mentioned earlier, the only satisfactory controls for the solid particles, gases and other airborne contaminants carried in the city's air are institutional and technical. Much improved air-conditions have been achieved in Great Britain due to tight air pollution control laws and regulations. At the same time it is evident that a return to the use of coal would again aggravate the air pollution problem. There is also little doubt that air pollution from vehicular emissions will continue to exist for many years as a reality of urban climate.

Where air pollution is dilute, however, an important environmental control device is plants. It has long been known that plants filter dust in cities. Ongoing research in plant physiology suggests that they do more than act as filters. The surface area of a tree has evolved to maximize light and gas exchange. Trees have ten times the surface area of the soil on which they stand. A hundred-year-old beech, for instance, has been estimated to have some 800,000 leaves, and a leaf surface area of 16,000 square metres per 160 square metres of tree base. Calculations show that the intercellular spaces of leaves (the sum of cell walls) increase the total leaf area to roughly 160,000 square metres.[47] Leaves can take up or absorb pollutants such as ozone and sulphur dioxide to significant levels. Urban vegetation can mitigate ozone pollution by lowering city temperatures and directly absorbing the gas.[48] It has been demonstrated that a Douglas fir with a diameter of 38 centimetres can remove 19.7 kilograms of sulphur dioxide per year without injury from an atmospheric concentration of 0.25 parts per million (p.p.m.). By way of illustration of the effectiveness of trees in removing sulphur dioxide, it shows that to take up the 462,000 tonnes of sulphur dioxide released annually in St Louis, Missouri, it would require 50 million trees. These would occupy about 5 per cent of the city's land area.[49]

Measurements taken in 1962 in Hyde Park in London, indicated that the concentrations of sulphur dioxide were reduced, partly because of the local circulation of air generated by the vegetation, and partly because of the uptake of gas by the leaves.[50]

Soil micro-organisms are more effective than vegetation in removing carbon monoxide and assist in the conversion of carbon monoxide to carbon dioxide. It is believed that oxygen released by roadside plants may help in lowering carbon monoxide levels along heavy traffic routes. Nitrogen oxide combined with gases such as oxygen produces nitrogen dioxide, which is then readily absorbed by vegetation.[51]

Vegetation also collects heavy metals. In New Haven, Connecticut, one researcher found that a sugar maple 30 centimetres in diameter removed 60 milligrams (mg) of cadmium and 140 mg of lead from the atmosphere during one growing season.[52] This suggests that vegetated spaces can provide areas where dust can settle out and where air pollutants are diluted. However, it is evident that plant damage occurs when pollutants are excessive. Schmid points out that the severity of plant damage is complicated by many factors, such as the age of the plant, its state of nutrition, moisture when exposed and other factors.[53] Plant species also vary in their tolerance to air pollution and their effectiveness in improving air quality. In effect, plants cannot be regarded as the panacea for ameliorating air pollution problems, but they do assist air purity and serve one other important climatic function – as indicators of air pollution and thus of the health of the people who live in cities. There are, therefore, highly valid reasons for the reforestation of urban areas in the planning and design of the urban environment.

Energy

While the larger issues of air quality control are beyond the scope of this inquiry, it is now conventional wisdom that a major threat to human, and non-human health in cities derives from vehicular emissions. An analysis of the energy efficiencies of various transportation options in terms of the amount of energy used in British Thermal Units (BTUs) to transport a passenger 1.6 kilometres in Canada demonstrates that the more efficient freight transportation modes are the most frequently used. However, in passenger transportation, the most heavily used mode – the car – is the least energy efficient. This is true for both urban and inter-city transportation.[54] A report on the future of urbanization by the Worldwatch Institute has shown that inner-city residents of New York use only one-third of the gasoline of residents living in the outer regions of the tri-state metropolitan area of New York, New Jersey and Connecticut; Manhattan residents use on average only 400 litres of gasoline per capita each year, a consumption level close to European cities.[55] The relationship between gasoline consumption and urban density in major cities around the world is illustrated in Table 6.1 and clearly

Table 6.1 *Urban density and gasoline consumption in major cities, 1980*

City	Gasoline consumption per person (megajoules)	Overall population density (People per hectare)	Inter-city population density (People per hectare)	Overall job density (Jobs per hectare)	Inner-city job density (Jobs per hectare)	Private automobile travel (per person, kilometres)
American cities[1]	58,541	14	45	7	30	12,507
San Francisco	55,365	16	59	8	48	13,200
Chicago	48,246	18	54	8	26	11,122
Australian cities[1]	29,849	14	24	6	27	10,680
Melbourne	29,104	16	29	6	40	10,128
Sydney	27,986	18	39	8	39	9,450
Metro Toronto	34,813	40	57	20	38	9,850
European cities[1]	13,280	54	91	31	79	5,595
Frankfurt	16,093	54	63	43	74	6,810
Stockholm	15,574	51	58	34	62	6,570
Paris	14,091	48	106	22	60	4,199
Asian cities[1]	5,493	160	464	71	296	1,799

Notes: 1 The figures given for American, Australian, European and Asian cities in the table are average for the cities in those regions studied by Newman and Kenworthy. The data reflect results from ten American, five Australian and twelve European cities. The Asian data are for the three 'westernized' Asian cities of Tokyo, Singapore and Hong Kong

2 *Interpretation:* The extent to which people choose to use private passenger cars is a direct function of urban density. Cities which are more compact are significantly less dependent on automobiles. Residents of US cities drive their cars, on average, two to three times as far as do residents of European cities. Toronto is about mid-way between a typical American and a typical European city in this regard

Source: Robert Paehlke, 'The Environmental Effects of Intensification', prepared for Municipal Planning Policy Branch, Ministry of Municipal Affairs, 1991. Based on Canadian Urban Institute, *Housing Intensification: Policies, Constraints and Challenges*, Toronto: Canadian Urban Institute, 1990, p. 13. Compiled from material originally presented in Peter Newman and Jeffrey Kenworthy, *Cities and Automobile Dependence: An International Sourcebook*, Hampshire: Gower Publishing, 1989

demonstrates the implications of current urban form on energy use. For instance, Houston, Texas, with approximately eight persons per hectare uses four and a half times as much gasoline as Copenhagen with a density of thirty persons per hectare. And at the extreme end of the scale, Houston's gasoline consumption is over five times that of Hong Kong with 300 persons per hectare.[56]

Thus gasoline consumption and air quality are essentially linked to urban form in the low-density suburban and fringe areas of the city, and reinforce the dominant role of the automobile in western society.[57] It also has radical implications for such issues as the consumption of agricultural land, damage to habitat, biological diversity and depletion

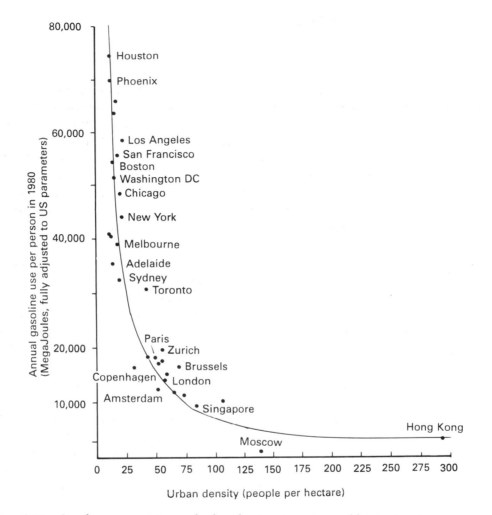

Figure 6.11 *Gasoline consumption and urban densities in major world cities, 1980*

The relationship between increasing densities and reducing energy consumption is clearly evident

Note: *Figures for all cities have been adjusted to US income, vehicle efficiency and gasoline prices*

Source: *Robert Paehlke, 'The Environmental Effects of Intensification', prepared for Municipal Planning Policy Branch, Ministry of Municipal Affairs, 1991, after Peter Newman and Jeffrey Kenworthy,* Cities and Automobile Dependence: An International Sourcebook, *Hampshire: Gower Publishing, 1989*

of non-renewable energy sources, and the need for producing alternative and more compact urban development patterns. One way of resolving these issues is illustrated by the case of Davis, California, discussed on pp. 262–3 and 279–83.

IMPLICATIONS FOR DESIGN

Green lungs

We have found that water and plants are the important natural elements of climate amelioration in the city. But what is their sphere of influence? How much vegetation or surface area of water is needed to have a marked effect on the climate of the city? Where should they be located? Answers to these questions depend on the climate of the region, the nature and variations of climate between one place an another, the characteristics of the site, its topography and the nature of the built-up area. At a small scale, experience tells us that a sheltered, well treed environment is a cooler and more pleasant place on a hot day. The sphere of influence of its elements – tree canopy, shrubs and water – would appear to be local, however, particularly if it is an isolated place in the general fabric of streets and buildings. The walled gardens of old European towns create a sphere of influence within them that are considerably cooler, or warmer, than outside. The impact of major spaces, the so-called 'green lungs' of the city, that have long been the ideal of landscape planning, may well be limited in the overall urban climate. Jane Jacobs has observed that the term 'green lung' is only applicable to the park spaces themselves and has little effect on the overall quality of the air in the city.[58] There have been claims that the oxygen produced by vegetation can affect the balance of oxygen in the air. These have evolved from the fact that much more oxygen is produced in photosynthesis than is used up in respiration. There is, however, as much oxygen used up in plant decay and the metabolism of animals that feed on plants as there is released by photosynthesis. The concept of the 'green lung' may be inaccurate in describing the effect of parks on the oxygen content of the city overall, but research has established definite connections between forest vegetation cover, open space distribution and urban climate control.

Meiss observes that the necessary amount of open space in an urban area and its optimum distribution cannot be stated quantitatively.[59] But from a climatic point of view, a fine mesh of small spaces, distributed evenly over the whole city, is more effective that reliance on a few large ones. These latter spaces need to be supplemented by a large number of small parks throughout the built-up area. 'Such a mesh facilitates horizontal exchange of air bodies of varying temperatures and consequently a balance is reached more quickly and with less resistance.'[60] A study of Dallas and Fort Worth found that the heat island reached its peak not in central areas of tall buildings, but along the fringes of the downtown area that contained low buildings and parking lots.[61] This suggests that efforts to ameliorate climate should be concentrated in these areas. Bernatzky suggests that the effect of the heat island can be partially counteracted by concentric rings of vegetated space to filter and oxygenate the air as it moves inward to the city centre.[62] In summer the heat from increasing densities in the city centre heats air which rises, creating a low air pressure area. This in turn draws cooler air in from the edges of the city. The quality of the air as it moves into the city is increasingly degraded, accumulating gases and particulate matter and being progressively de-oxygenated. Parks and vegetated spaces located in the path of this moving air will alter wind flow, ameliorate air quality and reduce temperature. In Chicago, air flow modelling has concluded that a finger plan with corridors of development and wedges of open space would have the most positive effect on air quality.[63]

Some design principles

There is still much to be learned about the effects of these natural processes on the city-wide climate, and how the scientific data that have been accumulated may be applied in determining optimum patterns of open space. Researchers in the field have varying views on what those patterns might be. We must, however, be careful to avoid the trap that often awaits those seeking answers to such problems – the temptation to create cookie-pattern solutions for every urban situation. This, in fact, is precisely the criticism that has been levelled at many planning theories in the past that have attempted to seek standard solutions for the cities. They ignore the inherent individuality and uniqueness of each place and, therefore, of each city. Thus, while there can be no final or definitive solution to the questions that have been explored here, certain general principles do emerge that are both pertinent and useful to this inquiry.

- The natural patterns of the land, its mountains, hills and valleys, its rivers, streams, open water, forest and grassland, determine local climatic patterns and affect, in some measure, the environment of the city. Although the extent of this influence may be local, the retention and enhancement of natural features for climatic reasons, are essential parts of the design process. An example of where climatic patterns, conditioned by topography and vegetation, have had a major impact on city form is the case of Stuttgart, Germany, described on pp. 278–81.
- Vegetation and water have a major effect on the maintenance of an equable micro-climate within cities. Since the large areas of paved and hard surfaces in the city generate the greatest heat in summer, establishing canopy vegetation wherever possible will reduce the adverse effects of the urban heat island. It will also remove dust and purify toxic gases and other chemicals. Dense canopies that provide maximum shade are much more effective than current practice, which sees trees as individual specimens. In addition to micro-climatic benefits, the continuity of the tree canopy also provides connected environments for some species of wildlife. In South Africa many parking areas are shaded by canvas canopies, to protect vehicles from

the summer sun, and also as a protection against frequent and damaging hailstorms. The retention of water and natural ponds in parks and other locations is also critical to restoring the energy balance by direct evaporation, and they serve many other hydrological and wildlife functions which have discussed in previous chapters.

- The large roofscape areas of downtown and suburban industrial sites contribute to heat build-up. The development of rooftop planting serves a similarly important role in climate amelioration. There is a need for basic practical research into lightweight low-maintenance techniques for establishing plants on existing rooftops, similar to the work in Germany described on p. 261. The use of naturalized urban vegetation that can survive with little or no care, and under the most severe conditions, has important implications for climate control.

- In most successful examples of climate manipulation in extreme conditions, the emphasis has been on an urban texture of small spaces and low buildings. In hot summer climates, narrow streets and small living spaces increase shade and reduce the build-up of solar radiation. They are, in addition, easier to control artificially through the use of orientation, shaded canopies, arcades, plants, water and ventilation. In cooler climates they are less subject to cold winds, drifting snow and extremes of temperature. Organization of space can create suntraps by orienting buildings to the south and excluding winter winds. In most cities where built form has evolved with little regard for climatic considerations, the materials of the landscape, as well as new built form, must serve a climatic role. Plants, groundform and building canopies, must be used to create suntraps and sheltered places, to counteract prevailing winds, modify down draught and so on.

The response to urban climate is the first step in the establishment of a vernacular; of a regional character that links built form with the place in which it occurs. Without this response, the designed landscape may be as clearly identified with the now defunct International Style as was the architectural style that gave rise to the name, but which none the less continues under different titles. Several contemporary examples of form evolving from climatic determinants are illustrated in the following pages.

Example 1 A northern university

In the late 1960s and early 1970s the University of Alberta underwent a major expansion programme to accommodate almost a doubling of its student population. Located within the city of Edmonton, Alberta, the university is very much a part of an urban environment that had one of the fastest growth rates in Canada. One of the most critical objectives of the development plan was to create an appropriate campus environment for living and working, since the period of greatest activity in the academic year takes place during the severe winter months.

Over the years the evolution of the campus had been dictated by conventional, or haphazard, planning. Buildings were placed formally in the landscape, isolated from each other by large open areas, and reflecting prevailing attitudes and nostalgia for wide open spaces and a rural prairie setting. The inappropriate nature of this model for the climate and environment of the prairies was shown by the way people moved around the campus; cutting through buildings not designed for the purpose in an effort to avoid the unpleasant conditions created by unimpeded winds, gusting around tall structures, and permanent shade cast by ill-placed structures. An examination of building coverage revealed that less than 15 per cent of the campus was occupied by buildings. The remaining 85 per cent was taken up by parking lots, roads, service areas, manicured turf and residual space that made much of the campus unusable and inhospitable.

An examination of climatic data will indicate that Edmonton is situated 53.35° north in a region of cold temperate climate. The coldest time of the year, and the period of least sunshine (100 hours and less per month between November and January), occurs during the academic year. While prevailing winds are from the south, the highest velocities occur from the north-west. During these periods, the combination of wind and cold temperatures creates, on occasion, a wind-chill factor of more than −45°C, enough to freeze flesh in less than 1 minute. In addition, sun angles at the winter solstice (22 December) are only 13 to 15° at noon. The two most critical design factors in cold climates, then, are wind and sun. The absence of wind and the presence of sunshine in the shaping of spaces can make the outdoors a very pleasant experience even when temperatures are very low. The development plan for the university, therefore, incorporated the following climatic criteria which became the guiding principles for its growth.

- The acquisition of additional land for campus expansion along conventional lines was rejected in favour of a policy of infilling within existing campus boundaries. This increased land coverage by buildings from 15 per cent to 34 per cent and greatly increased the efficiency and economy of land use. Part of the increased coverage took the form of parking structures on many levels to reduce the area taken up by surface parking and to increase usable and contained outdoor spaces. Infill buildings incorporated interior pedestrian streets linking academic departments, housing, food and recreation functions that provided alternative sheltered pedestrian routes during the winter months. The concept of infill and linkage also reinforced the notion of a high mix of uses, accessibility to services and the social integration that isolated university buildings tend to discourage.[64]
- Since the greatest need was to maintain maximum winter sunlight, development was designed to avoid tall structures on the south side of outdoor spaces. Alternatively

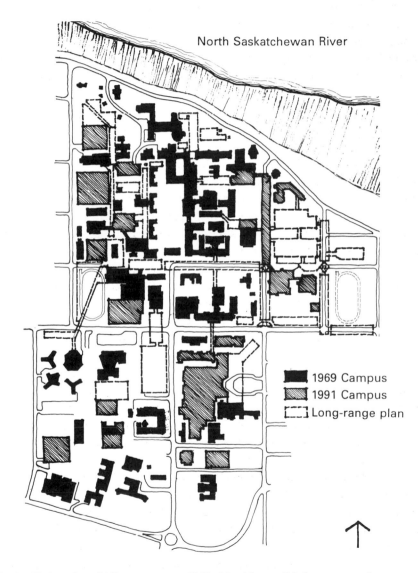

North Saskatchewan River

■ 1969 Campus
▨ 1991 Campus
⌐⌐ Long-range plan

Figure 6.12 *University of Alberta campus 1969–91 with possible long-range plan*

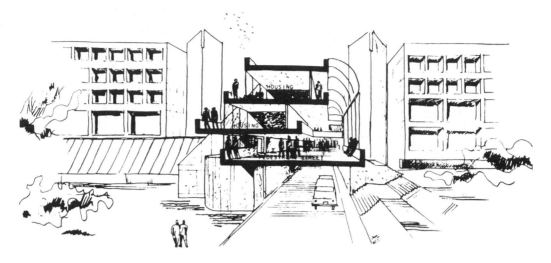

Figure 6.13 *Design principles. Connecting interior street and vehicular/pedestrian separation with housing over*

Source: *A.J. Diamond and Barton Myers, 'University of Alberta, Long Range Development Plan', consultant report, June 1969*

indoor/outdoor activity areas could be located on the south side of buildings.
- A limit of four to six storeys was recommended for most new building to reduce down draughts and gusting and increase available sunlight in winter.

These planning principles set a framework for outdoor space that could respond to basic design principles for climatic control at a micro-climatic scale. They included:

- the creation of small, well defined, protected courtyards and places for passive use, densely planted to create protected winter environments;
- the use of raised landforms, screens and planting around entrances, sitting and meeting places, to reduce wind chill and create suntraps;
- the use of coniferous wind breaks along pedestrian routes and surrounding large spaces such as athletic areas;
- dense planting or screens at narrow building openings and connected spaces where venturi winds are normally generated;
- reduction of glare from snow and creation of visual interest by the addition of plants and sculptural elements in the winter landscape;
- deciduous tree canopies that provide shade and temperature reduction over paved surfaces in the heat of the summer while allowing full exposure to the winter sun.

Plate 6.5 *Courtyard associated with the Student Union Building in January*

Smaller spaces and dense vegetation reduce wind and make conditions quite pleasant on a sunny day

Plate 6.6 *The same Student Union Building courtyard in May*

(Photo: Andrew Beddingfield)

Plate 6.7 *Major walkway protected by vegetation and buildings*

(Photo: Andrew Beddingfield)

Plate 6.8 *Student Union Building*

Linked buildings provide interior routes during inclement weather. Outdoor areas function as alternative routes and as gathering places year round

(Photo: Andrew Beddingfield)

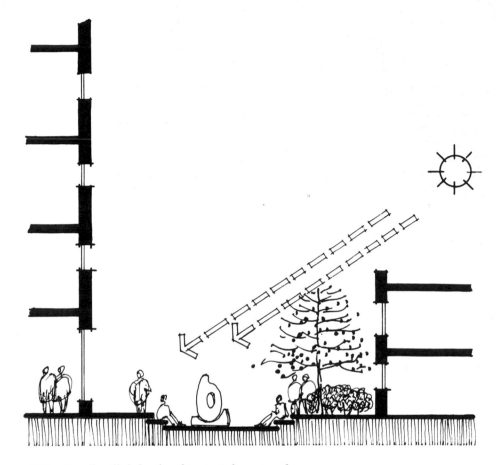

Figure 6.14 *Small well-defined and protected courtyards*

Dense deciduous canopies also contribute to the filtering of wind from above where a tall building creates down draughts.[65]

Since the 1970s, further phases of the long-range plan have been completed; the densely planted spaces have matured and the climatic strategies for the newly formed courtyards have begun to fulfil essential functions, creating benign winter and summer placcs, active university spaces and the experiential qualities needed for a true response to the climate of the region. The university has begun to develop an urban quality that reflects the climatic environment of this prairie region.

Figure 6.15 *Walkways protected by windbreaks in large open areas*

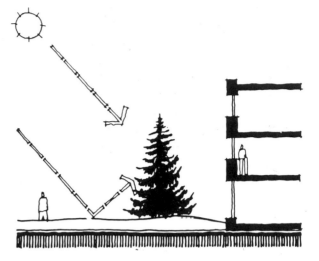

Figure 6.16 *Snow glare and intense light from low sun may be reduced by the use of coniferous or dense deciduous plants*

Example 2 Climate and city form

The links between major landforms, vegetation and urban form are nowhere better seen than in the city of Stuttgart. Located in the centre of an industrial region or over 2 million people, the city was plagued by air pollution from industrial stacks, cars and the use of coal and oil for heating dwellings, a situation greatly accentuated by frequent temperature inversions due to its location. The city occupies two valleys lying at right angles to each other. The main part of the city is situated in a basin-like valley (the Nesenbach), which is surrounded on three sides by steep slopes which extend on ridges into the centre of the city. The remainder is located along the open valley of the Neckar. The pollution problem became worse when urban expansion started to occupy the valley sides, replacing vineyards and forests with buildings. These buildings interrupted the normal katabatic flows of air, associated with valley land formations, from the surrounding vegetated hills to the city which had originally acted to minimize air inversions. It became apparent that there were interrelated links between open space distribution, climatic phenomena generated by topography and wind flows and a healthy city environment.

The prohibition of coal and oil for home heating was included in 150 local plans in the Stuttgart region and replaced with natural gas, reducing sulphur dioxide emissions by over 100 tonnes per year.[66] A network of parks was established in the city, linked to each other, to the river, forests and vineyards on the valley slopes. The forests, productive commercial resources, the vineyards and other arable lands account for about two-thirds of the total municipal area of 20,723 hectares.[67] Climatically, the vineyards tend to be more efficient than recreational lands since the vertical rows of vines permit uninterrupted air flow down valley sides. The function and character of the city's spaces is highly varied. Stuttgart's natural topography is such that the city's central parks are flat and those further out are located on steep valley sides that provide breezes down wooded slopes and a dramatic setting within the urban area. Within the parks themselves (which cover some 490 hectares), the use of plants, lakes and water sculpture has been exploited to provide places full of refreshing and varied sights and sounds. Green spaces are on the average 3°C cooler than surrounding built-up areas.[68] On both a city-wide and human scale, the parks and working landscapes within and surrounding Stuttgart are among the most climatically functional, socially useful and aesthetically pleasing of any modern city in the western world. From a landscape planning perspective their influence on its climate is matched by their function as 'green infrastructure', giving form and identity to the city.

Example 3   Energy conservation policies: Davis, California

Davis, California, is a city that has become internationally recognized for its integrated approach to energy conservation, one that has linked energy-efficient building, community design (including street and lot layout, landscape design and city-wide landscape planning), transportation, wildlife habitat, recycling of materials, water conservation, gardening and neighbourhood social life.

Davis' energy-conscious general plan was adopted in the early 1970s. It was the result of intensive background technical research by a number of working committees, and widespread public involvement. This in turn led to the development of the final Energy Conservation Code in 1975, unanimously adopted by the City Council the same year.[69] Climatic studies carried out on the thermal performance of buildings found that some buildings became dangerously hot due to direct solar heat gains through large east- or west-facing windows, while identical buildings with north- or south-facing windows remained comfortably cool and used substantially less energy for cooling. Similarly, south-facing windows exposed to the winter sun were significantly warmer during the winter – over 6°C on cold sunny days.[70] As part of this study the Davis climate was examined in light of the needs for energy conservation. Daytime maximum temperatures during July, the hottest time of the year, average 36°C. The night-time minimum, however, averages 13°C, a phenomenon caused by thermally induced sea breezes

Figure 6.17 *View of Stuttgart from forested slopes*

Figure 6.18 *Connections link park system together*

Figure 6.19 *Water forms a major element in the parks*

Figure 6.20 *Plants and shade over paved surfaces are distributed everywhere in Stuttgart*

originating over the Pacific Ocean and flowing into portions of the Central Valley. These findings led to the adoption of minimum performance standards for heat loss and gain during the winter and summer, which were consequently reflected in the more appropriate orientation of residential buildings, street widths and landscaping. City policies encouraged smaller lot sizes, greater development densities and the protection of agricultural land. A reduction of automobile travel time in favour of bicycles led to reduced energy use.[71]

Davis' plan also recognized the need to knit existing open spaces together with future growth areas, and called for a reconnection of the city with its larger regional landscape. In this regard, the plan created more than 80 kilometres of connected trails and public accessways, and is the first large-scale open space preservation plan in the United States that includes both riparian habitats and agricultural lands that range in scale from the family farm to corporate agribusiness.[72]

In a 1992 document entitled 'Energy Efficient Subdivision Design',[73] planners for the City of Davis identified certain general plan policies as being relevant to the review of development projects: those that encourage energy-efficient subdivision and building design, and those that require site planning to maximize the effects of south-west cooling winds through the design review process.

These policies continue to be used to implement city codes on energy conservation and solar access. Interpretation of basic policies (approved by the City Council in July 1992) show how they have integrated energy-efficient design with a wide range of considerations. Among these are: the preservation of natural features; the siting of streets, lots, buildings and street trees; the provision of mixed use and facilities within walking distance; and the arrangement of higher density and intensified uses near community facilities. In addition, this document includes guidelines on overall neighbourhood design, and design criteria that should be used in the review of subdivisions and planned developments. These have to do with the protection of natural and human-made features, response to climate, drainage systems that minimize impacts on city storm drainage systems and contribute to the underground water regime, pedestrian and bicycle circulation and other related aspects of design.[74] Ordinance No. 1618, for instance, sets out criteria for water conserving and drought tolerant landscapes that require at least 90 per cent of plant materials selected to be well suited to the climate of the region.[75] In addition, Davis' conservation programme recognizes the importance of trees as a natural air-conditioner, since neighbourhoods with shaded narrow streets can be as much as 10°C cooler than those with unshaded streets and use only use half of the amount of electricity for air-conditioning.[76]

Many of these policies, modified and improved over the years, are embodied in 'Village Homes', an innovating 242 home subdivision located on 28 hectares of land. This development addressed issues that many Davis residents were adopting – alternative

Plate 6.9 *Village Homes in Davis, California, on a rainy winter's day*

Backyards and accessways are designed to retain rainwater during and after storms. This project shows great attention to detail. Note change in materials where a walkway crosses a drainage swale

modes of transportation, energy efficiency and community interaction.[77] Designed and built by Michael Corbett in the early 1970s, the development includes detention ponds for rainwater, a variety of orchards, community gardens, vineyards and common areas dispersed throughout the neighbourhood, narrow 6 or 7 metre wide cul-de-sac streets, pedestrian and cycle greenways and a south-facing layout for lots and houses. The development also includes an abundant landscape and street trees, many of which are fruit-bearing. A post-occupancy evaluation undertaken in August 1990 compared Village Homes with a standard control subdivision. It concluded, among other factors, that:

- home energy use and energy consumption for transportation purposes are approximately one-third less than in conventional subdivisions, attributable to solar design of homes, less air-conditioning and less use of vehicles;
- residents grow more fruit and vegetables which contribute approximately 25 per cent of the household's annual consumption of these commodities;
- recycling behaviour has been found to be generally higher than in the control subdivision;
- residents appreciate the social attributes of Village Homes, such as semi-public spaces, a community centre and appropriate places for children and social contact.
- residents socialize much more that their counterparts in the control subdivision, and know their neighbours better.[78]

SOME CONCLUDING REFLECTIONS ON URBAN CLIMATE

One may be tempted to comment that the value of plants, landform and water to the creation of beneficial micro-environments has been 'rediscovered' by the contemporary science of climatology. It has begun to measure something that people knew by trial and error and have made use of for generations in pre-industrial built environments. Applying traditional methods of climate control, in effect, now has the blessing of the scientific method.

But there are climatic phenomena that the older cities were rarely, if ever, faced with. Atmospheric pollution, urban heat islands and down draughts from tall buildings, drainage systems and hydrological imbalances are a creation of large industrial cities. Creating favourable habitats by natural means combines traditional wisdom, modern science and intelligent planning. The use of plants and water on the walls, floor and roof surfaces of the city can create natural climatic control and can in large measure restore the energy balance through evaporation of water into the air and the metabolic processes of plants. The arrangement of built and landscape elements can reduce the impact of wind, take advantage of sun and create favourable micro-climatic environments.

Following the principle of economy of means, the natural patterns and materials of the landscape can be made to work for the city environment in new ways and at less cost, compared to the energy costs currently incurred to maintain highly inhospitable conditions. The question 'how can human development *contribute* to the environments it changes?', provides us with a positive and proactive basis for action that can restore healthy climates, biological diversity and productive urban soils. It also provides us with a foundation for internalizing nature in human affairs. And if the natural and human processes that sustain life are integrated and visibly parts of everyday life, we enrich immeasurably the urban environment.

It is evident that climatic forces form a common thread that influences all the other natural and human processes of water, plants, animals, urban farming and city life. Together they are an integration of natural and human systems and interconnected elements that play a vital role in the ecological health and quality of life in cities, and, in the process, provide the multiple opportunities and benefits that we have been examining.

•

In this and previous chapters I have tried to show how, by bringing urban and natural processes together at a local level, a new design language emerges that has significance for the evolving form of the city. The forces that have shaped today's urban regions have been governed by unlimited energy resources and by attitudes that have paid little heed to the necessity for a sustainable future, or to the relevance of nature to cities. The urban environment has been managed on a piecemeal basis, treating the economy separately from social issues, or the environment; consequently, solutions to problems have been single-minded, simplistic and fragmented. The pieces of the puzzle have remained separated to the detriment of natural systems and the quality of urban life as an enriching social, environmental and sensory experience.

It has also become evident that the climate of cities must be understood in a wider perspective that crosses regional boundaries, since the influences of solar energy, global wind patterns, the atmosphere, fresh waters, the oceans and continents and human activities on the land, link cities together in a worldwide interconnected whole. Air and water pollution, global warming, depletion of the ozone layer, vanishing forests, poverty and hunger, once thought of as having local consequences, or as being problems that belong somewhere else, now become much debated global issues affecting everyone, and crossing all political boundaries.

Some important and far-reaching shifts in values are beginning to emerge, however, that have to do with notions of renewal, citizen empowerment and action, co-operation as opposed to confrontation between people and governments, and the integration of nature, economy and social agendas in decision-making. Many of the approaches to achieving a sustainable future for cities that may have seemed unworkable or unrealistic in the 1960s are being recognized as achievable and necessary in the last decade of the twentieth century.

Per Stig Moller, the Danish Minister of the Environment wrote in 1991 that

> Economic development must go hand in hand with re-creation of the ecological balance. We must not look on environmental pollution as an isolated problem. This is the vital point that we must always remember when we debate our social and economic problems.[79]

This perspective is increasingly recognized in other political arenas as a central determinant for change. In effect, a long-term investment in things like soil fertility, perpetuation of forests and environmental stability may be seen as a better bargain than quick profits gained at the expense of the future. The practical application of the principles outlined in this book depends less on altruistic motives, and more on what makes practical sense. Alternative ways of doing things are eventually done for pragmatic reasons, because the alternatives provide tangible benefits, in ecological, economic or social terms. Many of the examples I have illustrated are occurring fortuitously under our noses, but are now being recognized for what they are, and for the lessons they can teach. Where necessity prevails, such as in times of war, or the perennial problems of poverty, the need to grow food, recycle materials or engender a sense of community purpose will continue to be a matter of survival. In the 1990s, society has begun to recognize that global or local environmental sustainability will be determined largely by our cities. Such a goal will be dependent on finding ways to protect natural wealth, regenerate the processes of urban nature and adopt more environmentally conscious lifestyles.

The idea of interdependence between human communities and nature, enshrined in Barry Commoner's well known principle that 'everything is connected to everything else', illustrates that the more one learns about one's home place, the more relevant this becomes to the larger global community, and the greater the need to protect biological diversity beyond local boundaries. Global thinking must become a framework for local action, since the two are inextricably linked. The key to environmental sanity and civilized urban life in the twenty-first century may well lie in how we deal with these interrelated realities.

NOTES

•

INTRODUCTION

1 Boardman, Philip. *The Worlds of Patrick Geddes*. London: Routledge and Kegan Paul, 1978.

1 URBAN ECOLOGY: A BASIS FOR SHAPING CITIES

1 World Commission on Environment and Development. *Our Common Future*. Oxford and New York: Oxford University Press, 1987.

2 I have borrowed the word 'pedigreed' from Bernard Rudofsky's *Architecture Without Architects*. New York: Museum of Modern Art, 1964. Rudofsky uses it to describe the formal architecture of cities that express power and wealth: in this case the formal designed urban landscape.

3 Mumford, Lewis. *The City in History*. New York: Harcourt, Brace and World, 1961.

4 Ibid.

5 Rudofsky. *Architecture Without Architects*.

6 Havlick, Spenser W. *The Urban Organism*. New York: Macmillan, 1974.

7 Giedion, Sigfried. *Space, Time and Architecture*. Oxford: Oxford University Press, 1952.

8 Laurie, Michael. 'Nature and City Planning in the Nineteenth Century'. In Ian C. Laurie (ed.). *Nature in Cities*. New York: John Wiley, 1979.

9 McHarg, Ian L. *Design with Nature*. Garden City, NY: Natural History Press, 1969.

10 Lang, Reg and Armour, Audrey. *Environmental Planning Resource Book*. Montreal: Lands Directorate, Environment Canada, 1980.

11 Steadman, Philip. *Energy, Environment and Building*. Cambridge: Cambridge University Press, 1977.

12 Loenthal, David. 'Daniel Boone is Dead'. *Natural History*. American Geographical Society, Aug.–Sept. 1968.

13 Ibid.

14 *Jakarta Post*, 12 May 1992.

15 Odum, Eugene P. 'The Strategy of Ecosystem Development'. *Science*, vol. 164, April 1969.

16 Commoner, Barry. *The Closing Circle*. New York: Knopf, 1971.

17 Royal Commission on the Future of the Toronto Waterfront. *Watershed*. Interim

Report, August 1990. Hon. David Crombie, Commissioner, Toronto.

18 Attributed to John A. Livingston, a member of our university group.

19 Fowles, John. 'Seeing Nature Whole'. *Harpers*, vol. 259, no. 1554, Nov. 1979.

20 Ibid.

21 Deelstra, Prof. Tjeerd. 'Enforcing Environmental Urban Management – New Strategies and Approaches'. Conference material for UN Conference on Environment and Development (UNCED). Berlin, 6 Feb. 1992.

22 Ibid.

23 Ibid.

2 WATER

1 Bellamy, David. *Bellamy's Europe*. London: British Broadcasting Corporation, 1976.

2 Environment Canada. *Water – Nature's Most Versatile Substance*. Ottawa: Inland Waters Directorate, 1976.

3 Ibid.

4 Petawawa Forest Experiment Station. *Water Trail*. Chalk River, Ontario: Public Awareness Program, Canadian Forest Service, 1978.

5 Golding, D.L. 'Forests and Water'. Fact Sheet. Ottawa: Department of Fisheries and Environment, n.d.

6 Ibid.

7 Hough Stansbury and Associates. 'Water Quality and Recreational Use of Inland Lakes'. Prepared for the Ontario Ministry of the Environment, SE Region, May 1977.

8 Vallentyne, J.R. *The Algal Bowl*. Ottawa: Department of the Environment, Fisheries and Marine Service, Miscellaneous Special Publication 22, 1974.

9 Royal Commission on the Future of the Toronto Waterfront. *Watershed*. Interim Report. Toronto, Aug. 1990.

10 Lull, Howard W. and Sopper, William E. *Hydrologic Effects from Urbanization of Forested Watersheds in the NE*. Washington DC: USDA Forest Service Research Paper NE 146, US Department of Agriculture, 1969.

11 Ontario Ministry of the Environment. *Evaluation of the Magnitude and Significance of Pollution Loadings from Urban Stormwater Run-off in Ontario*. Research Report no. 81.

12 Leopold, Luna B. *Hydrology for Urban Land Planning. A Guide Book on the Hydrologic Effects of Urban Land Use*. Geological Survey circular 554. Washington DC: US Department of the Interior, 1968.

13 Ontario Ministry of the Environment. *Modern Concepts in Urban Drainage*. Conference Proceedings no. 5, A Canada, Ontario Agreement on Great Lakes Water Quality, Toronto, 1977.

14 Ibid.

15 Lull and Sopper. *Hydrologic Effects.*

16 Forbes, R.J. 'Mesopotamian and Egyptian Technology'. In Melvin Kranzberg and Carroll W. Pursell Jr (eds). *Technology in Western Civilization*, vol. 1. Oxford: Oxford University Press, 1967.

17 Drachmann, A.G. 'The Classical Civilization'. In Kranzberg and Pursell (eds). *Technology in Western Civilization*, vol. 1.

18 Mumford, Lewis. *The City in History.* New York: Harcourt, Brace and World, 1961.

19 Wright, Lawrence. *Clean and Decent.* London: Routledge and Kegan Paul, 1960.

20 Mathews, Leslie S. *The Antiques of Perfume.* London: G. Bell, 1973.

21 Wright. *Clean and Decent.*

22 Mumford. *The City in History.*

23 Goldstein, Jerome. *Sensible Sludge.* Emmaus, Pa.: Rodale Press, 1977.

24 Ibid.

25 Wood, L.B. 'The Rehabilitation of the Tidal River Thames'. Unpublished Thames Water Authority Paper, n.d.

26 Ibid.

27 Royal Commission on the Future of the Toronto Waterfront. *Watershed.*

28 Holub, Hugh A. 'Water Conservation and Wealth: A Tale of Two Cities'. Salt River Project Waterline Spring 1987.

29 Deelstra, Prof. Tjeerd. 'Enforcing Environmental Urban Management – New Strategies and Approaches'. Conference material for UN Conference of Environment and Development (UNCED). Berlin, 6 Feb. 1992.

30 Wood. 'The Rehabilitation of the Tidal River Thames'.

31 Vallentyne. *The Algal Bowl.*

32 Task Force to Bring Back the Don. *Bringing Back the Don.* Toronto: Hough Stansbury Woodland, Prime Consultants, in association with, Gore and Storrie Ltd, Dr Robert Newbury, The Kirkland Partnership, 1991.

33 Task Force to Bring Back the Don. *Bringing Back the Don.*

34 Dr Robert Newbury, personal communication.

35 Task Force to Bring Back the Don. *Bringing Back the Don.*

36 Ibid.

37 Furedy, C. 'Wastes and the Urban Environment: Perspectives on People, Animals, and Their Wastes'. Calcutta: Institution of Public Health Engineers, India, with WEDC Loughborough University of Technology and British Deputy High Commissioner, 1985.

38 Sopper, William E. *Surface Application of Sewage Effluent and Sludge.* Madison, Wisc.: ASA-CSSA-SSSA, 1979.

39 Feigin, A., Ravina., I. and Shalhevet, J. *Irrigation with Treated Sewage Effluent,*

Management for Environmental Protection.
Berlin: Springer-Verlag, 1991. Reported in
Kenneth, R. Tamminga. 'Land Application of
Treated Municipal Effluent: Literature Review
of Post-Implementation Evaluation'.
Department of Landscape Architecture,
Pennsylvania State University, University Park,
Pa. n.d.

40 Hills, Lawrence K. 'Putting Waste Water to
Work'. *Ecologist*, vol. 5, no. 9, 1975.

41 Sopper. *Surface Application.*

42 Ibid.

43 Tamminga. 'Land Application of Treated
Municipal Effluent'.

44 Baylin, Frank. 'Solar Sewage Treatment'.
Popular Science, May 1979.

45 Burke, William K. 'From Sewer to Swamp'.
E Magazine, July–Aug. 1991, pp. 39–42 and 67.

46 Carroll, Debra. 'Greenhouses that Grow
Clean Water'. *Sunworld*, vol. 14, no. 3, 1990,
pp. 71–4.

47 Ibid.

48 Burke. 'From Sewer to Swamp'.

49 Cited in Burke. 'From Sewer to Swamp'.

50 Deelstra, Tjeerd. 'Ecological Approaches to
Wastewater Management in Urban Regions in
the Netherlands'. In *Ecological Engineering for
Wastewater Treatment*. Proceedings of the
International Conference, Stensund Folk
College, Sweden, March 1991.

51 Ibid.

52 Ontario Ministry of the Environment.
*Stormwater Quality Best Management
Practices*. Toronto: Queen's Printer for Ontario,
1991.

53 Hough, Michael. *The Urban Landscape*.
Toronto: Conservation Council of Ontario,
1971.

54 Mark E. Taylor and Associates. *Constructed
Wetlands for Stormwater Management: A
Review*. Report prepared for Water Resources
Branch, Ontario Ministry of the Environment
and Metropolitan Toronto and Region
Conservation Authority, April 1992.

55 Ibid.

56 Bunyard, Peter. 'Sewage Treatment in a
Swedish Sculpture Garden'. *Ecologist*, no. 1,
Jan.–Feb. 1978.

57 Hess, Alan. 'Technology Exposed'.
Landscape Architecture, vol. 82, no. 5, May
1992.

58 Wallace, McHarg, Roberts and Todd were
the environmental planning consultants in a
multi-disciplinary team hired in 1971 to prepare
plans for the site.

59 McHarg, Ian L. and Sutton, Jonathan.
'Ecological Plumbing for the Texas Coastal
Plain'. In Grady Clay (ed.). *Water and the
Landscape*. New York: McGraw-Hill, 1979.

60 Coyle, Tom. 'The Application of Ecological
Planning to the Urban Planning and
Development Process'. Unpublished student

paper, Faculty of Environmental Studies, York University, Toronto, 1980.

61 Hough Stansbury and Associates. 'LeBreton Flats Landscape Development'. Unpublished report, LeBreton Flats Office, Central Mortgage and Housing Corporation, Ottawa, Jan. 1979.

3 PLANTS

1 Odum, Eugene P. *Ecology.* New York: Holt, Rinehart and Winston, 1963.

2 Odum, Eugene P. 'The Strategy of Ecosystem Development'. *Science*, vol. 164, April 1969.

3 Farb, Peter. *The Forest.* New York: Time-Life Books, 1963.

4 Elias, Thomas S. and Irwin, Howard S. 'Urban Trees'. *Scientific American*, Nov. 1976.

5 Jorgensen, Erik. 'Forestry: Some Problems and Proposals'. Faculty of Forestry, University of Toronto, Sept. 1967.

6 Laurie, Michael. 'Nature and City Planning in the Nineteenth Century'. In Ian C. Laurie (ed.). *Nature in Cities.* New York: John Wiley, 1979.

7 Fairbrother, Nan. *New Lives, New Landscapes.* London: Architectural Press, 1970.

8 Sukopp, Herbert, Blume, Hans-Peter and Kunick, Wolfram. 'The Soil, Flora and Vegetation of Berlin's Wastelands'. In Laurie. *Nature in Cities.*

9 Livingston, John A. *Canada.* Toronto: Jack McClelland, 1970.

10 Sukopp, Blume and Kunick. 'Soil, Flora and Vegetation'.

11 Bellamy, David. *Bellamy's Europe.* London: British Broadcasting Corporation, 1976.

12 Hammond, Herb. *Seeing the Forest Among the Trees.* Vancouver: Polestar, 1991.

13 Teagle, W.G. *The Endless Village.* Shrewsbury: Nature Conservancy Council, 1978.

14 Sukopp, Blume and Kunick. 'Soil, Flora and Vegetation'.

15 Dorney, R.F. and Eagles, Paul F.J. *et al.* 'Ecosystem Planning, Analysis and Design in Ontario as Applied to Environmentally Sensitive Areas'. Paper to American Association for the Advancement of Science Meeting. Toronto, Jan. 1981.

16 Hough, M. and Barrett, Suzanne. *People and City Landscapes.* Toronto: Conservation Council of Ontario, 1987.

17 Jorgensen, 'Forestry'.

18 Guldemond, J.L. 'City Woodlands'. Talk to Department of Town and Country Planning, University of Manchester, Sept. 1978, unpublished.

19 Hough Stansbury Michalski Ltd. 'Naturalization Project'. Report. Ottawa: National Capital Commission, 1982; and,

'Naturalization Project: Five Year Test Plot Evaluation'. Ottawa: National Capital Commission, 1988.

20 Hough Stansbury Michalski Ltd. 'Naturalization Project'.

21 Hough Stansbury Woodland. 'Naturalization Project. Five Year Test Plot Evaluation'.

22 Ibid.

23 Ibid.

24 Bos, H.J. and Mol, J.L. 'The Dutch Example: Native Planting in Holland'. In Laurie. *Nature in Cities*.

25 Ibid.

26 Jacobs, Ton. 'The Gilles Estate in Delft'. In Tjeerd Deelstra (ed.). *Shaping Nature in Cities*. In Press.

27 Ibid.

28 Sybrand Tjallingi, personal communication.

29 Ruff, Allan R. *Holland and the Ecological Landscapes, 1973–1987*. Delft: Delft University Press, 1987.

30 Jacobs. 'The Gilles Estate in Delft'.

31 City of Toronto Planning and Development Department. 'A Working Guide for Planning and Designing Safer Urban Environments'. Toronto: Safe City Committee, City Hall Toronto, 1992. Quoted from Nottingham Safer Cities Project. *Community Safety in Nottingham City Centre: Report of the Steering Group*. Nottingham: Nottingham Safer Cities Project, Oct. 1990.

32 Ibid.

33 City of Toronto Planning and Development Department. 'A Working Guide'.

34 Emil Frolich, forest engineer, Stadtforstamt, Zurich, personal communication.

35 *New York Times*, 5 May 1991.

36 Pollan, Michael. 'Why Mow? The Case Against Lawns.' *New York Times Magazine*, 28 May 1989.

37 Ibid.

38 Ibid.

39 Ibid.

40 Morrison, W.O. *et al.* 'Avi Fauna Survey of Vacant Grasslands'. Ottawa: National Capital Commission, 1981.

41 Smith, R.A.H. and Bradshaw, A.D. 'Use of Metal Tolerant Plant Populations for the Reclamation of Metalliferous Wastes'. *Applied Ecology*, no. 16, 1979.

42 Cole, Lyndis. 'Conservancy in Urban Areas'. Report for the Nature Conservancy Council. Land Use Consultants, May 1978.

43 Lowday, J.E., and Wells, T.C.E. *The Management of Grassland and Heathland in Country Parks*. Report to the Countryside Commission by the Institute of Terrestrial Ecology. Shrewsbury: Countryside Commission, West Midlands Region, 1977.

44 Laurie, Ian C. 'Urban Commons'. In Laurie. *Nature in Cities*.

45 Weaver Liquifuels, Downsview, Ontario, personal communication.

46 Thomsen, Charles, Department of Landscape Architecture, University of Manitoba, personal communication.

47 Hough Stansbury and Associates. 'LeBreton Flats Landscape Development'. Unpublished report for Central Mortgage and Housing Corporation, Ottawa, Jan. 1979.

48 Jacobs, Jane. *The Life and Death of Great American Cities*. New York: Random House, 1961.

49 Brower, Sidney. 'Streetfront and Sidewalk'. *Landscape Architecture*, July 1973.

50 Francis, Mark. 'The Making of Democratic Streets'. In Anne Vernez Moudon (ed.). *Public Streets for Public Use*. New York: Van Nostrand Reinhold, 1987.

51 Ibid.

52 Ibid.

53 Murray, Alex. 'User Control as a Design Parameter of Urban Open Space'. Metamorphosis conference, Thessaloniki, Greece, May 1992.

54 Ibid.

55 Hough Stansbury Woodland with Professor Alex Murray. 'Ataratiri Open Space Study'. City of Toronto Housing Department, June 1990.

56 *Globe and Mail*, Toronto, 16 Oct. 1991.

57 Grove, Noel. 'Greenways: Paths to the Future'. *National Geographic*, vol. 177, no. 6, June 1990.

58 Goulty, George. 'Landscape Electric'. *Landscape Design*, Aug. 1986.

59 Jones, Brad. 'Comparison of the Washington State and Ontario Regulatory Frameworks'. Paper in proceedings of the Lower Donlands: Site Remediation Workshop, Toronto, May 1993.

60 Schmid, Arno Sighart. 'The Interdisciplinary Significance of Water in the Ecological Renewal of the Emscher Region'. In Sybrand Tsallingi (ed.). *Hydropolis Reader. The Role of Water in Urban Planning*. Proceedings of an International Workshop, 29 March–2 April 1993. Wageningen, Netherlands.

61 'The Emscher Park International Building Exhibition. An Institution of the State of North-Rhine Westphalia'. Information Brochure, Sept. 1991

62 Arno Sighart Schmid, IBA Emscher Park, personal communication.

63 IBA Handbook. 'Emscher Park, Examples of Projects', June 1992.

64 Hough, Michael. *The Urban Landscape*. Toronto: Conservation Council of Ontario, 1971.

65 Whyte, William H. *The Last Landscape*. Garden City NY: Doubleday, 1968; Anchor edn, 1978.

66 Deelstra, Tjeerd. 'National, Regional and Local Planning Strategies for Urban Green Areas in the Netherlands. An Ecological Approach'. Seminario internacional sobre uso, tratamiento y gestion del Verde Urbano. Institut d'Ecologia Urbana de Barcelona. Barcelona, Nov. 1988.

67 Forman, R.T.T. and Godron, M. *Landscape Ecology.* New York: John Wiley and Sons, 1986.

68 Hough Stansbury Woodland/Gore and Storrie Ltd. 'Ecological Analysis of the Greenbelt'. Ottawa: National Capital Commission, 1991.

69 Ibid.

4 WILDLIFE

1 Belloc, Hilaire. *The Bad Child's Book of Beasts, Together with More Beasts for Worse Children.* London: Duckworth, 1974.

2 Livingston, John A. *Rogue Primate.* Toronto: Key Porter Books Ltd, 1994.

3 Farb, Peter. *The Forest.* New York: Time-Life Books, 1963.

4 Ibid.

5 Tofts, Richard and Clements, David. 'Hedgerows as Wildlife Habitat'. *Landscape Design Extra*, Jan. 1993.

6 Geis, Aelred K. 'Effects of Urbanization and Types of Urban Development on Bird Populations'. In *Wildlife in an Urbanizing Environment Symposium.* Co-operative Extension Service, University of Massachusetts, Amherst, June 1974.

7 Wilcove, D.S. 'Forest Fragmentation as a Wildlife Management Issue in the Eastern United States'. Quoted in Hough Stansbury Woodland/Gore and Storrie Ltd. 'Ecological Analysis of the Greenbelt'. Ottawa: National Capital Commission, 1991.

8 Urban Wildlife Research Centre. *Urban Wildlife News*, March 1979.

9 Livingston, John A. *Canada.* Toronto: Natural Science of Canada, 1970.

10 Ibid.

11 Douglas, Gena (ed.). *The William Curtis Ecological Park.* Second Report. London: Ecological Parks Trust, 1979.

12 Sukopp, Herbert. 'Urban Ecosystems'. *J. Nat. Hist. Mus. Inst.*, Chiba, vol. 2, no. 1, pp. 53–62, March 1992.

13 Barrett, Suzanne and Kidd, Joanna. *Pathways: Towards an Ecosystem Approach.* Toronto: Royal Commission on the Future of the Toronto Waterfront, 1991.

14 Bonnett, Penelope A. 'London: An Examination of the City as a Wildlife Area'. In Gerald McKeating (ed.). *Nature and Urban Man.* Canadian Nature Federation Conference, University of Western Ontario, 1974.

15 Karstad, Lars. 'Disease Problems of Urban Wildlife'. In David Euler *et al.* (eds). *Wildlife in Urban Canada, Symposium*. Guelph, Ontario: University of Guelph, 1975.

16 Ibid.

17 More, Thomas A. 'An Analysis of Wildlife in Children's Stories'. In *Children, Nature and the Urban Environment*. Symposium proceedings. Washington DC: USDA Forest Service, General Technical Report, NE 30, US Department of Agriculture, 1977.

18 Worster, Donald. *Nature's Economy.* Garden City, NY: Anchor Press/Doubleday, 1979.

19 Livingston. *Rogue Primate.*

20 Forman, Richard, T.T. and Godron, Michel. *Landscape Ecology.* New York: John Wiley, 1986.

21 Livingston. *Canada.*

22 Hamel, Peter, J. 'Wastewater Treatment and Shoreline Ecology in the Regional Municipality of Ottawa-Carlton'. Unpublished student paper, Faculty of Environmental Studies, York University, Toronto 1976.

23 Ibid.

24 Catling, Paul M., McIntosh, Karen L. and McKay, Sheila M. 'The Vascular Plants of the Leslie Street Headland'. *Ontario Field Biology,* vol. 31, no. 1, 1977.

25 Ibid.

26 Metropolitan Toronto and Region Conservation Authority. *Tommy Thompson Park, Master Plan and Environmental Assessment.* Addendum report, Dec. 1992.

27 Blokpoel, Hans and Haymes, Gerard T. 'How the Birds Took Over the Leslie Street Spit'. *Canadian Geographic Journal*, April–May 1978.

28 Metropolitan Toronto and Region Conservation Authority. 'Tommy Thompson Park Newsletter'. n.d.

29 Blokpoel and Haymes. 'How the Birds Took Over the Leslie Street Spit'.

30 The Friends of the Spit. 'The Eastern Headland and Aquatic Park'. Produced by the Friends of the Spit, Toronto, unpublished, n.d.

31 The Friends of the Spit. 'Brief to the MTRCA re Addendum', Dec. 1992; 'Tommy Thompson Park Master Plan and Environmental Assessment', 25 June 1993.

32 Ibid., emphasis in the original.

33 *Toronto Globe and Mail*, 12 June 1993.

34 Johnson, Bob. *Familiar Amphibians and Reptiles of Ontario.* Toronto: Natural Heritage/Natural History Inc., 1989.

35 Ibid.

36 Dougan, J. 'The fate of ESA's in Urban Environments: Two Case Histories in Peel and Halton'. *Plant Press* vol. 2 nos 7–9, 1984. Quoted in Hough Stansbury Woodland/Gore and Storrie Ltd. 'Lake Ontario Greenway Strategy: Restoration Ecology'. Unpublished report. Toronto: Waterfront Regeneration Trust, May 1994.

37 Hough Stansbury Woodland/Gore and Storrie Ltd. 'Lake Ontario Greenway Strategy'.

38 Gore and Storrie Ltd. 'Forest Wildlife Monitoring Program, Year 2'. Prepared for Ontario Hydro, Land Use and Environmental Planning Department, 1991. Quoted in Hough Stansbury Woodland/Gore and Storrie Ltd. 'Ecological Analysis of the Greenbelt'.

39 Robinson, Steve. 'Environmental Implications of Golf Courses'. *Landscape Design Extra*, April 1992.

40 Tiner, Tim. 'Green Space or Green Waste?' *Seasons*, Federation of Ontario Naturalists, summer 1991.

41 Etchells, Jon. 'Golf Answers Back'. *Landscape Design*, no. 210, May 1992.

42 Baines, Chris. *How To Make a Wildlife Garden*. London: Elm Tree Books, 1985.

43 Ibid., p. 189.

44 Hamel. 'Wastewater Treatment and Shoreline Ecology'.

45 Hough Stansbury and Associates. 'Public Awareness Program'. Petawawa National Forest Institute, Canadian Forestry Service, Ottawa, 1972.

46 *Groot Haags Groenboek*. The Hague: Hague Municipal Public Relations Bureau Publication, 1971.

47 O'Connor, Glenn A. 'Establishing and Maintaining Low Maintenance Landscapes for Southern Ontario'. Unpublished research report prepared for the University of Toronto, Faculty of Architecture and Landscape Architecture, 1984.

48 Jim Laut, Environmental Officer, Lakeshore Refinery, Mississauga, personal communication.

49 Karstad. 'Disease Problems'.

50 Southwood, T.R.E. 'The Number of Species of Insects Associated with Various Trees'. *Journal of Animal Ecology*, vol. 30, 1961.

51 Bellamy, David. *Bellamy's Europe*. London: British Broadcasting Corporation, 1976.

52 Vanstone, Ellen. 'Raccoons. The Bandit in the Attic'. *Toronto Globe and Mail*, 20 Feb. 1993.

53 Ibid.

54 Barrett and Kidd. *Pathways: Towards an Ecosystem Approach*.

55 Duffey, Eric. 'Effects of Industrial Air Pollution on Wildlife'. *Biological Conservation*, vol. 15, 1979, pp. 181–90.

56 Ibid.

57 Ibid.

5 CITY FARMING

1 Hopkinson, Frank N. 'Country Code, Unauthorized Version'. *Farmer's Weekly* (UK), Feb. 1980.

2 Heady, Earl O. 'The Agriculture of the U.S'. *Scientific American*, issue of Food and Agriculture, Sept. 1976.

3 Commoner, Barry. *The Closing Circle*. New York: Knopf, 1971.

4 Geno, Barbara J. and Geno, Larry M. *Food Production in the Canadian Environment*. Perceptions 3. Ottawa: Science Council of Canada, 1976.

5 The New Alchemist Institute. 'Modern Agriculture: A Wasteland Technology'. *Journal of the New Alchemists*, 1974.

6 Sinclair, Geoffrey. 'Upland Landscape Study'. Unpublished address to the Employment in the Countryside Symposium, New Mills Study Centre, Devon, May 1980.

7 Feder, Natasha. *Population and Global Sustainability*. Toronto: Conservation Council of Ontario, 1992.

8 Cited in Brown, Lester R. and Jacobson, Jodi L. *Our Demographically Divided World*. Worldwatch Paper 74. Washington DC: Worldwatch Institute, Dec. 1986.

9 Green, John. 'Memorandum on Domestic Livestock Keeping in Urban Areas'. Unpublished report, 1943.

10 Animal Control Sub-Committee. 'Proposals for Animal Control'. City of Toronto, 1981.

11 Mumford, Lewis. *The City in History*. New York: Harcourt, Brace and World, 1961.

12 Ibid.

13 Mitchell, W.R. 'The Cow Keepers'. *The Dalesman*, vol. 43, no. 11, 1980.

14 *Indonesia Times*, 15 May 1992.

15 Ibid.

16 Sachs, Ignacy and Silk, Dana. *Food and Energy: Strategies for Sustainable Development*. Publication based on activities of the Food-Energy Nexus Programme, 1983–8. Paris: United Nations University, 1989.

17 Ibid.

18 Feder, Natasha. *Population and Global Sustainability*. Toronto: Conservation Council of Ontario, May 1992.

19 Hill, Lawrence D. *Fertility Without Fertilizers*. Braintree, Essex: Henry Doubleday Research Association, 1975.

20 Green. 'Memorandum on Domestic Livestock Keeping'.

21 Ibid.

22 John Green. personal communication.

23 Harrowsmith Report. 'The City Farmers', *Harrowsmith*, vol. 3, no. 20, July 1979, p. 8.

24 *USA Today*, international edition. 14 August 1993.

25 Ibid.

26 Ibid.

27 Wade, Isabel. 'Urban Self-Reliance in the Third World: Developing Strategies for Food and Fuel Production'. World Futures Conference, Toronto, 1980.

28 Ibid.

29 Memon, Pyar Ali and Lee-Smith, Diana. 'Urban Agriculture in Kenya'. CJAS/RCEA, vol. 27, no. 1, 1993.

30 *Toronto Star*, 12 Feb. 1994.

31 Furedy, C. 'Wastes and the Urban Environment: Perspectives on People, Animals and Their Wastes'. Calcutta: Institution of Public Health Engineers, India with WEDC, Loughborough University of Technology and British Deputy High Commission, 1985.

32 Ibid.

33 Ibid.

34 *Jakarta Post*, 12 May 1992.

35 Ibid.

36 Yudhi Komarudin, Corporate Secretary, Bata Shoes, personal communication.

37 Poerbo, Hasan. 'Urban Solid Waste Management in Bandung: Towards an Integrated Resource Recovery System'. *Environment and Urbanization: Rethinking Local Government – Views from the Third World*, vol. 3, no. 1, April 1991, pp. 60–9.

38 Ibid.

39 *Toronto Globe and Mail*, 25 Oct. 1991.

40 *Toronto Globe and Mail*, 9 Dec. 1992.

41 Brown, Lester R. *et al. State of the World 1992*. New York: Worldwatch Institute, 1992.

42 Thompson, J. William. 'San Francisco's Gardens of Diversity'. *Landscape Architecture*, vol. 83, no. 1, Jan. 1993.

43 Thompson, J. William. 'Reconsidering South Central'. *Landscape Architecture*, vol. 83, no. 1, Jan. 1993.

44 Cited in Thompson. 'Gardens of Diversity'.

45 Cited in Ibid.

46 The Farallones Institute. *The Integral Urban House. Self-Reliant Living in the City*. San Francisco: Sierra Club Books, 1978.

47 Hough, Michael. 'Bottom-Line Horticulture'. *Harrowsmith*, no. 56, Aug.–Sept. 1984.

48 Unpublished calculations by the author.

49 Riley. *Economic Growth*.

50 Wagner, Judith Joan. 'The Economic Development Potential of Urban Agriculture at the Community Scale'. Unpublished Master of City Planning Thesis, Massachusetts Institute of Technology, 1980.

51 Hayes, Denis. *Energy. The Case for Conservation*. Worldwatch Paper 4,

Washington DC: Worldwatch Institute, Jan. 1976.

52 BBC Television. *Tomorrow's World*, 21 Feb. 1980.

53 Pollock, Cynthia. *Mining Urban Wastes: The Potential for Recycling*. Worldwatch Paper 76, Washington DC: Worldwatch Institute, April 1987, p. 15.

54 Rifkin, Jeremy (ed.). *The Green Lifestyle Handbook*. New York: Henry Holt and Co., 1990.

55 Ibid.

56 Ibid.

57 Mumford. *The City in History.*

58 Lumley-Smith, Ruth. 'The Road to Utopia'. *New Ecologist*, no. 1, Jan.–Feb. 1978.

59 *Toronto Star*, 20 Nov. 1979.

60 Memon and Lee-Smith. 'Urban Agriculture in Kenya'.

61 Hough, Michael and Barrett, Suzanne.

People and City Landscapes. Toronto: Conservation Council of Ontario, 1987.

62 Creasy, Rosalind. *The Complete Book of Edible Landscaping*. San Francisco: Sierra Club Books, 1982.

63 Mollison, Bill and David Holmgren. *Permaculture I*. Condell Park, New South Wales: Corgi, 1978.

64 *The California Aggie*, 9 May 1985, p. 1.

65 Thompson. 'Gardens of Diversity'.

66 Laurie, Michael. 'Nature and City Planning in the Nineteenth Century'. In Ian C. Laurie (ed.). *Nature in Cities*. New York: John Wiley, 1979.

67 Cited in Fein, Albert. *Landscape into Cityscape*. Ithaca, NY: Cornell University Press, 1968.

68 Local Economic Development Information Service (LEDIS). Initiative A465, July/Aug. 1992.

69 Ibid.

6 CLIMATE: MAKING CONNECTIONS

1 Graham, Harry. *Ruthless Rhymes for Heartless Homes*. London: Edward Arnold, reissued 1974.

2 Meiss, Michael. 'The Climate of Cities'. In Ian C. Laurie (ed.). *Nature in Cities*. New York: John Wiley, 1979.

3 Chandler, T.J. *The Climate of London*. London: Hutchinson, 1975.

4 Miess, 'The Climate of Cities'.

5 Ford Foundation. *Exploring Energy Choices; A Preliminary Report of the Ford*

Foundation's Energy Policy Project.
Washington DC: The Ford Foundation, 1974.

6 Ibid.

7 Ibid.

8 McPherson, E. Gregory, Nowak, David J. *et al. Chicago's Evolving Urban Forest: Initial Report of the Chicago Urban Forest Climate Project.* Northeastern Forest Experiment Station, US Department of Agriculture, Forest Service. General Technical Report NE-169, 1993.

9 Landsburg, H.E. 'The Climate of Towns'. In William L. Thomas Jr (ed.). *Man's Role in Changing the Face of the Earth*, vol. 2. Chicago: University of Chicago Press, 1956.

10 Illinois Environmental Protection Agency 1991. *Illinois 1990 Annual Air Quality Report.* IEPA/APC/91–92, Springfield, Illinois. Reported in McPherson, Nowak *et al. Chicago's Evolving Urban Forest.*

11 *Guardian*, 26 March 1993.

12 Oke, T.R. *Boundary Layer Climates*. New York: Methuen, 1987. Quoted in McPherson, Nowak *et al. Chicago's Evolving Urban Forest.*

13 Lowry, William P. 'The Climate of Cities'. *Scientific American*, Aug. 1967.

14 Akbari, H. *et al. The Urban Heat Island: Causes and Impacts. In Cooling Our Communities: A Guidebook on Tree Planting and Light-Coloured Surfacing.* Washington DC: US Environmental Protection Agency. Quoted in McPherson, Nowak *et al. Chicago's Evolving Urban Forest.*

15 McPherson, Nowak *et al. Chicago's Evolving Urban Forest.*

16 Burke, James. *Connections*. London: Macmillan, 1978.

17 Rotsch. Melvin M. 'The Home Environment'. In Melvin Kranzberg and Carroll W. Pursell Jr (eds). *Technology in Western Civilization*, vol. 2. Oxford: Oxford University Press, 1967.

18 Carter, George F. *Man and the Land, a Cultural Geography*. New York: Holt, Reinhart and Winston, 1966.

19 Cain, Allan, Afshar, Farroukh *et al.* 'Traditional Cooling Systems in the Third World'. *Ecologist*, vol. 6, no. 2, 1976.

20 Ibid.

21 Ibid.

22 Rudofsky, *Architecture Without Architects*. New York: The Museum of Modern Art, 1964.

23 Bahadori, Mehdi N. 'Passive Cooling Systems in Iranian Architecture', *Scientific American*, Feb. 1978.

24 Turner, T.H.D. 'The Design of Open Space'. In Timothy Cochrane and Jane Brown (eds). *Landscape Design for the Middle East*. London: RIBA Publications, 1978.

25 Robinette, Gary O. *Landscape Planning for Energy Conservation*. Reston, Va.: Environmental Design Press for the American Society of Landscape Architects Foundation, 1977.

26 Ibid.

27 Meiss. 'The Climate of Cities'.

28 Doernach, Rudolf. 'Über den Nutzen von biotektonischen Grunsystemen'. *Garten + Landschaft*, no. 6, 1979.

29 Ibid.

30 Falk, Trillitzsch. 'Anregungen zum Thema Dachgarten'. *Garten + Landschaft*, no. 6, 1979.

31 Federer, C.A. 'Effect of Trees in Modifying Urban Microclimate'. In *Trees and Forests in an Urbanizing Environment Symposium*. Amherst: Co-operative Extension Service, University of Massachusetts, 1970.

32 Ibid.

33 Ibid.

34 Schmid, James A. 'Urban Vegetation'. University of Chicago, Department of Geography, Research Paper no. 161, 1975.

35 Philleo, Jerryne. *The Davis Energy Handbook*. Davis: City of Davis, 1981.

36 Ibid.

37 Nowak, D.J. 'Urban Forest Structure and the Functions of Hydrocarbon Emissions and Carbon Storage'. In McPherson, Nowak *et al. Chicago's Evolving Urban Forest*.

38 Akbari, H. 'The Impact of Summer Heat Islands on Cooling Energy Consumption and CO_2 Emissions'. In McPherson, Nowak *et al. Chicago's Evolving Urban Forest*.

39 Tucker, John. 'Trees for People'. *Landscape Design*, no. 215, Nov. 1992, pp. 41–3.

40 Meiss, 'The Climate of Cities'.

41 Oke, T.R. 'The Significance of the Atmosphere in Planning Human Settlements'. In E.B. Wiken and G. Ironside (eds). *Ecological (Biophysical) Land Classification in Urban Areas*. Ecological Land Classification Series, no. 3. Ottawa: Environment Canada, 1977.

42 Robinette, *Landscape Planning*.

43 Olgay, Victor G. *Design with Climate*. Princeton: Princeton University Press, 1963.

44 Robinette, *Landscape Planning*.

45 McPherson, Nowak *et al. Chicago's Evolving Urban Forest*.

46 Ontario Ministry of Housing. *Residential Site Design and Energy Conservation, Part 1, General Report*. Toronto: Government of Ontario, 1980.

47 Salvatore, Fidenzio. *The Potential Role of Vegetation in Improving the Urban Air Quality, A Study of Preventative Medicine*. Willowdale, Ontario: Mork-Toronto Lung Association, 1982.

48 McPherson, Nowak *et al. Chicago's Evolving Urban Forest*.

49 Schmid. 'Urban Vegetation'.

50 Wainwright, C.W.K. and Wilson, M.J.G. 'Atmospheric Pollution in a London Park'. *International Journal of Air and Water Pollution*, vol. 6, 1962.

51 Salvatore, *The Potential Role of Vegetation.*

52 Schmid. 'Urban Vegetation'.

53 Ibid.

54 Energy Probe. 'Facts on Energy Conservation'. Toronto: Energy Probe, n.d.

55 Brown, Lester and Jacobson, Jodi L. *The Future of Urbanization: Facing the Ecological and Economic Constraints.* Washington DC: Worldwatch Institute, 1987.

56 Paehlke, Robert. 'The Environmental Effects of Intensification'. Prepared for Municipal Planning Policy Branch, Ministry of Municipal Affairs, 1991. Quoted from 'Canadian Urban Institute Housing Intensification, Policies Constraints and Challenges'. Background paper. Toronto 1: Canadian Urban Institute, 1990.

57 Paehlke. 'The Environmental Effects'.

58 Jacobs, Jane. *The Life and Death of Great American Cities.* New York: Random House, 1961.

59 Meiss. 'The Climate of Cities'.

60 Ibid.

61 Schmid. 'Urban Vegetation'.

62 Bernatzky, Aloys. 'The Effect of Trees on the Climate of Towns'. In Shirley E. Wright *et al.* (eds). *Tree Growth in the Landscape.* Department of Horticulture, Wye College, University of London, 1974.

63 Salvatore. *The Potential Role of Vegetation.*

64 Diamond, A.J. and Myers, Barton. 'University of Alberta, Long Range Development Plan', consultant report, June 1969.

65 Hough Stansbury and Associates. 'University of Alberta, Long Range Landscape Development Plan', consultant report, June 1971.

66 Baumuller, J. 'The Infrastructure of Climatology in the Administration and Development of Planning in the City of Stuttgart'. In Oke, T.R. (ed.). *Urban Climatology and its Applications with Special Regard to Tropical Areas.* Geneva: Secretariat of the World Meteorological Organization, 1986.

67 International Federation of Landscape Architects, *10 Kongress der Internationalen Federation der Landschaftarchitekten.* Stuttgart, 1966.

68 Cox, Jenny. 'The Green Ways of Stuttgart'. *Landscape Design*, no. 110, 1975.

69 Philleo. *The Davis Energy Handbook.*

70 Ibid.

71 Ibid.

72 Francis, Mark, Dawson, Kerry and Jones, Stan. *The Davis Greenway Plan.* Davis: Centre for Design Research, June 1989.

73 City of Davis Community Development Department. 'Energy Efficient Subdivision Design'. Approved by Council, 29 July 1992.

74 Ibid.

75 Ordinance No. 1618. An ordinance amending article XXIII of Chapter 29 of the Davis Municipal Code relating to water conservation standards for new construction landscaping, n.d.

76 Philleo. *The Davis Energy Handbook*.

77 Yee, Suzanne Joy. 'Davis: A Sustainable City'. Student thesis, University of California at Davis, Landscape Architecture Department, June 1992.

78 Lentz, Thomas. 'A Post-Occupancy Evaluation of Village Homes, Davis, CA'. Study undertaken as part of a master's degree in social and urban geography from the Technical University of Munich, Aug. 1990.

79 Cited in Jespersen, Soren Steen. 'Denmark, Innovator in Environmental Planning'. *Scandinavian Review*, vol. 79, no. 1, spring–summer 1991.

SELECT BIBLIOGRAPHY

•

This bibliography is a small selection of references that has been compiled with two purposes in mind: to provide further readings on the various subjects the book covers; and to provide up to date reference material for those subjects where the literature has expanded significantly since the first edition was published.

FOUNDATIONS OF ENVIRONMENTAL THOUGHT

Evernden, Neil. *The Natural Alien*. Toronto: University of Toronto Press, 1985. The author contends that the environmental movement's ability to achieve its goals is in doubt, despite the publicity given environmental issues today. He reviews the inherent assumptions of western industrial societies; assumptions that work in opposition to environmental causes. The existence of differing approaches to human/environmental relations is, however, evidence of a basic characteristic of humanity – that of flexibility, and it is this that provides the basis for a new understanding of how as a species we relate to the non-human world. Only through this understanding, the author argues, can the impulse to environmental advocacy and crisis be understood and addressed.

Livingston, John A. *Rogue Primate: An Exploration of Human Domestication*. Toronto: Key Porter Books, 1994. This book challenges most conventional ideas about the relationship between people and the natural world. The author contends that the first domesticated animal was neither dog nor goat, but human. Humans cut themselves adrift from the real world by becoming entirely dependent on ideas, and technical ideas have provided our species with the power to manipulate nature. The natural world itself has consequently been drawn into the service of humans and their belief systems. Our understanding of nature is informed by an ideological insistence that domination is somehow 'natural', a social theory that lay behind Charles Darwin's explanation of evolution as the 'struggle for existence'. This book is necessary reading for anyone wishing to understand the impact of people on nature.

GENERAL READING

Boardman, Philip. *The Worlds of Patrick Geddes*. London: Routledge and Kegan Paul, 1978. A biography of Geddes' life, ideas, and achievements as a biologist, town planner, sociologist, educator. The book is an excellent introduction to this great pioneer of twentieth-century thought, and a precursor to Geddes' *Cities in Evolution* (1915) republished by Rutgers University Press (1972).

Hoskins, W.G. *English Landscapes*. London: British Broadcasting Corporation, 1979. This book is about the English landscape; how people have made their mark on the land over thousands of years. As the author says 'it has been written upon over and over again in a kind of code'. In chapters on settlement, colonization, boundaries, churches, transport and industry the author shows us how to read these immensely varied landscapes and understand how they came about.

Jacobs, Jane. *The Death and Life of Great American Cities*. New York: Random House, 1961 (latest edition). Jane Jacobs asks: what makes cities work? why are some neighbourhoods full of things to do and see and why are others dull? Over thirty years after its first publication, the observations of the city she made in the early 1960s are as relevant in the 1990s. She taught people to see places as they really are.

Laurie, Ian C. (ed.). *Nature in Cities*. New York: John Wiley, 1979. A series of informative chapters on urban climate, plants and urban ecosystems, nature and city planning in the nineteenth century, and alternative 'naturalized' approaches to parks and open spaces in modern cities. Remains useful and relevant in the 1990s.

McHarg, Ian L. *Design with Nature*. New York: Natural History Press, 1969 (second edition 1992). A seminal book that has had a major influence on modern environmental planning. McHarg's central theme is that of Ecological Determinism. The bio-physical processes that shape the physical landscape are deterministic; they respond to natural laws and are form-giving to human adaptations. Nature is a fundamental force that that will determine the morphology of modern cities and human endeavours.

Miller, Helen. *Patrick Geddes: Social Evolutionist and City Planner*. London: Routledge, 1990. Another excellent biography of Geddes with full bibliographic lists.

Mumford, Lewis. *The City in History*. New York: Harcourt, Brace and World, 1961. This classic exploration of the city – its origins, transformations and prospects – remains essential reading for understanding cities from all points of view.

National Greening Australia, Conference Proceedings. *A Vision for a Greener City: The Role of Vegetation in Urban Environments*. Freemantle, Western Australia: Greening Australia Limited, October 1994. (Available from Greening Australia Limited, GPO Box 9868 Canberra ACT 2601.) This publication examines problems and issues of Australia's Greening Program in relation to cities. It includes papers on the environment of urban areas, planning for conservation and development, management of urban environments with respect to vegetation, economic and social issues, and community involvement, and bio-diversity. Numerous case studies are included.

Papanek, Victor. *Design for the Real World*. New York: Pantheon Books, Random House, 1971 (second revised edition 1985). This book focuses on the integration of industrial design and natural processes. Its underlying message and themes, however, have direct application to environmental design and urban ecology. Useful reading.

Platt, Rutherford, H., Rowan, A. Rowntree and Muick, Pamela C. (eds). *The Ecological City: Preserving and Restoring Urban Biodiversity*. Amherst: The University of Massachusetts Press, 1994. A series of chapters on cities and natural environments. The book explores issues of geography, ecology, landscape architecture, urban forestry and environmental education, from broad overviews of problems to detailed case studies.

Royal Commission on the Future of the Toronto Waterfront. *Watershed*. Interim Report, August 1990. Obtainable from: Royal Commission on the Future of the Toronto Waterfront. 207 Queen's Quay West, 5th Floor, Toronto, Ontario. The first of a series of reports published by the Toronto Waterfront Commission. Outlines an integrative approach to planning based on the 'Ecosystem approach' that links between natural processes, people, human activities and the city. Suggests a series of principles for a green waterfront that flow from this approach and outlines prescriptions for regenerating the city.

Royal Commission on the Future of the Toronto Waterfront. *Regeneration: Toronto's Waterfront and the Sustainable City*. Final report, December 1991. The report summarizes the Commission's work over three years. Includes an analysis of the ecosystem approach; a suggested framework for ecosystem-based planning; and environmental imperatives that link water and the state of the Great Lakes, shoreline issues, greenways and the winter waterfront, healing an urban watershed (the story of the Don river), places within the Toronto bio-region, regeneration and strategies for recovery.

Thayer, Robert L. Jr. *Gray World, Green Heart: Technology, Nature, and the Sustainable Landscape*. New York: John Wiley and Sons, 1994. This book identifies three conflicting forces in each of us that the author believes lie at the heart of the environmental crisis: love of land and nature; dependence on and affection for technology; and fear of technology's negative side effects. They give rise to a sense of guilt that is expressed in the way we shape our living places; for instance, air conditioners hidden behind fences and bushes, or the use of consumptive technologies to build fake waterfalls. The author suggests an alternative theoretical framework upon which to build a future, one in which technologies support rather than dominate nature. Many concrete examples are given to illustrate the theory.

ON WATER

Stream hydrology/storm drainage/wastewater

Burke, William K. 'From Sewer to Swamp: the Green Answer to Dirty Water'. *E Magazine*, July/August 1991. An investigation into 'natural' sewage treatment through the use of aquatic plants and fauna in the USA. Discusses the need and applications of John Todd's approach, and the difficulties in gaining acceptance of such alternatives.

Deelstra, Tjeerd and Rob uit de Bosch. 'Ecological Approaches to Wastewater Management in Urban Regions in the Netherlands'. In *Ecological Engineering for Wastewater Treatment*. Proceedings of the

International Conference, 24–8 March 1991, Stensund Folk College, Sweden. This paper examines four aspects of wastewater management in the Netherlands: water purification by means of vegetated wetlands; encouragement by government of ecological engineering; financial comparisons of ecological and conventional systems; and how ecological engineering relates to planning.

Fretz, Laurie. *Urban Streams Rehabilitation Project* (report). Toronto: Conservation Council of Ontario, February 1992. An overview of the need for rehabilitation of urban streams, and an action plan that outlines a goal-oriented strategy that can be used by individuals, groups or agencies.

Gover, Nancy. 'Greenhouse Technology Treats Wastewater Naturally'. *Small Flows*, vol. 7 no. 2, April 1993. Another useful introduction to John Todd's work in solar aquatics. Describes the process, gives several examples and contact for further information.

Hongkong Standard. 'The Politics of Rivers'. Special section on global affairs prepared by the *Hongkong Standard*, November 1994. Discusses world issues of rivers, the dependency on rivers for existence in the middle east, how water is shared among nations, national security and related political problems. Interesting for its larger national focus in comparison to the watershed issues under discussion at more local levels in North America.

Hough Stansbury Woodland *et al. Bringing Back the Don*. Toronto: Task Force to Bring Back the Don, (City Hall, Toronto, Ontario), August 1991. The complete report on the restoration strategy for Toronto's Don river. Includes history, context and issues, potential strategies for restoration and a vision for the lower Don.

Marshall Macklin Monaghan Ltd. *Stormwater Management Practices: Planning and Design Manual*. Prepared for Ontario Ministry of Environment and Energy. June 1994. (Available from Ministry of Environment and Energy, Program Development Branch, Municipal Programs Section, 135 St Clair Ave. West, Toronto, Ontario, M4V 1PS.) A technical guide on the planning, design and review of stormwater management practices. It stresses an integrated approach that covers watershed planning to subdivision and site planning, management practices, and capital, operations and maintenance costs.

National Committees of the Netherlands and Germany for the International Hydrological Program of UNESO. 'Hydropolis Reader: the Role of Water in Urban Planning'. Proceedings of an international workshop on hydrology, held in Wageningen, The Netherlands, March 1993. A series of papers by workshop participants that focus on three themes: urban water systems, experiences and views of technicians and managers; water in urban development, experiences of planners and designers; the role of water in decision making and planning procedures. Useful case studies from the US and Canada, China, Australia, Ireland, Turkey and Europe. Contact: Section OCC (IAC) Box 88, 6700 AB Wageningen, The Netherlands.

Newbury, Robert W. and Marc N. Gaboury. *Stream Analysis and Fish Habitat Design*. Privately published by Newbury Hydraulics Ltd, Box 1173, Gibson's, British Columbia, 1993. A very useful technical field manual and guide to stream habitat projects. Includes planning, field exploration, evaluation of stream behaviour, design and construction of stream habitat works.

Tamminga, Kenneth R. 'Land application of Treated Municipal Effluent: Literature review of Post-implementation Evaluation'. Department of Landscape Architecture, Pennsylvania State University, University Park, PA 16802. Unpublished paper (n.d.). A selective survey and literature review of the overall environmental impacts and general efficiency of wastewater irrigation. Paper focuses on the infiltration capacity of soils and vegetation on sprayed, treated wastewater. It also addresses the larger framework of wastewater reuse for agricultural purposes. Includes the work of William Sopper (the Living Filter concept) and other field researchers in the USA and developing countries.

Water Planning Division, US Environmental Protection Agency. *Final Report of the Nationwide Urban Runoff Program*, Washington DC, 1983. A five-year project conducted by the US Environmental Protection Agency at 28 locations in the USA to characterize and evaluate urban runoff.

Wright, Lawrence. *Clean and Decent*. London: Routledge and Kegan Paul, 1960. A small but highly informative book on the social history of water throughout the ages, how it was used and disposed of, historical traditions in keeping clean, and the evolution of the bathroom.

ON PLANTS

(General/naturalization/restoration)

Antenen, Susan D. 'Forest Restoration Project in New York City. Wave Hill, Bronx, NY'. *Restoration and Management Notes*, vol. 4, no. 1, 1986. Restoration of a degraded urban woodland consisting of 3 ha of degraded woodland and 1 ha of exotic vines at Wave Hill, a public garden in the Bronx. Restoration efforts range from natural succession to planting of characteristic associations to increase diversity and create habitats.

Baines, C. and J. Smart. *A Guide to Habitat Creation*. London: The Greater London Council. 44 pp., 1984. This book highlights the creation of grasslands (meadows), woodlands and wetlands as urban environments for countryside plants and mammals. There are practical techniques for habitat creation (in Britain).

Bellamy, David. 'Doge City'. In *Bellamy's Europe*. London: British Broadcasting Corporation, 1976. This book was the product of a BBC television series and the chapter 'Doge City' is about the wide variety of plants one can find growing fortuitously in Venice. Very useful for a keen sense of observation and understanding of urban nature.

Berger, J.J. (ed.). *Environmental Restoration: Science and Strategies for Restoring the Earth.* Proceedings of conference on ecological restoration, Berkeley, CA, January. 1988. Washington, DC: Island Press, 398 pp., 1990. Scientific and technical papers on terrestrial restoration, aquatic restoration, law, planning, land acquisition and conflict resolution, all in connection to restoration.

Bouta, M. 'Restoring our Tallgrass Prairies'. *Am. Horticulturist*, no. 70, pp. 10–18, 1991. This author chronicles the work of Ray Schulenberg, Floyd Swink and Robert Betz as leaders of the prairie restoration movement in northern Illinois.

Bradshaw, A.D., D.A. Goode and E.H.P. Thorp (eds). 'Ecology and Design'. In Landscape, the 24th Symposium of the British Ecological Society. Manchester: Blackwell Scientific Publications, 463 pp., 1986. General reference work identifying issues relating to approach, design and management of natural areas in the European context. Covers natural approach in urban context.

Bragg, T.B. 'Implications for long-term prairie management from seasonal burning of loess hill and tallgrass prairies'. In Nodvin, S.C. and T.A. Waldrop (eds) *Fire and the Environment: Ecological and Cultural Perspectives; Proceedings of an International Symposium.* US Forest Service, southeastern forest experiment station, Asheville, NC, pp. 34–44, 1991. Managers should burn more areas of prairie during the growing season (summer and fall) in order to maintain and improve plant species diversity. These fires should not be so extensive as to jeopardize invertebrate populations.

Bray, Paul M. 'The City as a Park'. *American Land Forum Magazine*, winter 1985. The idea of a park as an oasis amidst the dense bricks and mortar of the city is thoroughly ingrained in our thinking. This thoughtful article suggests that all parts of the city – its work spaces, living quarters, streets and heritage, have equal aesthetic and recreational potential – qualities that are not just confined to the park, but must be seen as part of the entire structure of the urban environment.

Buckley, G.P. (ed.). *Biological Habitat Reconstruction.* London: Belhaven Press, 363 pp., 1989. Excellent reference on habitat reconstruction in England addressing principles of and opportunities for habitat reconstruction, the creation of new habitats and enhancing existing habitats. Both the urban and rural contexts are explored.

Forman, R.T.T. and M. Godron. *Landscape Ecology.* New York: John Wiley and Sons, 1986. This book on the study of the landscape is written from an ecologist's perspective, but is intended to be readily understandable to the non-scientist. It is based first, on direct observation of the visual landscape (its patterns and characteristics); second, by a scientific exploration of what made that landscape the way it is (its structure); third, how it behaves (its function). The book explores the natural science, artistic, social and economic dimensions of the landscape, that include terrestrial and aquatic environments, populations, ecological communities, energy flows and matter. Very helpful for a basic understanding of how landscapes work.

Hammond, Herb. *Seeing the Forest among the Trees: the Case for Holistic Forest Use*. Vancouver: Polestar Press, 1991. A valuable book for understanding forests as dynamic biological entities in relation to the practice and politics of forestry. The author discusses forest ecology and provides practical solutions to forestry problems in this context. Comprehensive and well illustrated, this book is important to an understanding of urban forests.

Harker, D. , S. Evans, M. Evans and K. Harker. *Landscape Restoration Handbook*. Boca Raton, FL: Lewis Publishers, 688 pp., 1993. A good reference book on landscape restoration with chapters on naturalizing the managed landscape, planning approaches for naturalizing the landscape, summary of ecological principles relating to biodiversity, conservation biology and ecosystem restoration, principles and practices of natural landscaping, bird and butterfly gardens, a detailed directory of natural regions in the United States and a listing of the plant communities with the primary woody and herbaceous species.

Hough, M. and Suzanne Barrett. *People and City Landscapes*. Toronto: Conservation Council of Ontario, 1987. A study of people and open space in metropolitan areas in Ontario. Deals with the special needs of different groups; different recreational activities and the types of open space that these require; and ecological design and management issues of different types of parkland.

Hough Stansbury Woodland. *Naturalization Project*. Ottawa: National Capital Commission, 170 pp., 1982. A review of the most economical techniques for restoring native woodland in urban areas. Alternative site preparation techniques, tree and shrub species and mulching methods tried on a number of test plots.

Hough Stansbury Woodland. *Naturalization Project: Five Year Test Plot Evaluation*. Ottawa: National Capital Commission, 106 pp., 1989. A follow-up study to review the success and failure of the various techniques employed in the 1982 Naturalization Project. Evaluation of the test plots, the survival of different species and the success of mulching techniques.

Hough Stansbury Woodland. *Urban Vegetation Management Study*. Ottawa: National Capital Commission, 73 pp., 1990. Practical techniques for slopes, water edges, wet and dry habitats, plant lists.

Hough Stansbury Woodland Naylor Dance, with Gore and Storrie Ltd. *Ecological Restoration Opportunities for the Lake Ontario Greenway*. Toronto: report prepared for the Waterfront Regeneration Trust, June 1994. Survey of ecological restoration and implementation strategies in practice as of 1994. Includes chapters on site selection, assessment of ecosystem functions, implementation techniques, vegetation communities and the urban landscape. Extensive annotated bibliography.

Kirt, R.B. 'Quantitative Trends in Progression toward a Prairie State by Use of Seed Broadcast and Seedling Transplant Methods'. In *Abstracts of the 12th N.A. Prairie Conference*, 1990. (see Smith, D.D.

and C.A. Jacobs (eds). 'Recapturing a Vanishing Heritage'. Proceedings of the twelfth North American Prairie Conference. University of Northern Iowa. Cedar Falls, IA, 1990). On the basis of a four-year study, results showed that prairie species could be transplanted equally well by broadcasting seeds or transplanting seedlings.

Kline, V.M. 'Henry Green's Remarkable Prairie.' *Restoration and Management Notes*, no. 10, pp. 36–7, 1992. Description of 20 ha prairie at the University of Wisconsin Arboretum. Planted by Dr Henry Greene, almost single-handed, between 1945 and 1953 using seeds, seedlings and wild transplants. Now a high quality area: high diversity of native species; dominance by native prairie grasses; relatively few exotic weed problems.

Lingenfelter, Bonita K. 'Butterfly Social'. *Harrowsmith*, no. 97, vol. xv 1:1, May/June 1991. This article discusses reasons for loss of habitats and species diversity of butterflies, and suggests ways of planting gardens to attract them. Design plans and a detailed list of flowering plants are included.

Luken, J.O. *Directing Ecological Succession*. New York: Chapman and Hall, 1990. Book concerned with the consequences of deliberately disturbing vegetation: to achieve a desired goal through manipulating the plant community by disturbance. Deals with methods of managing succession including plant and plant part removal, managing succession by changing resource availability, changing propagule availability and animals and succession.

North York, City 1985. *Naturalization Areas in North York*. North York: Parks and Recreation Department, 1985. A report outlining naturalization strategies to promote natural areas for recreation and environmental enhancement for twenty parks in the municipality of North York.

Potel, Sandra. *Reforesting the Earth*. Washington, DC: Worldwatch Institute, 1988. Many world-wide statistics and data on forest cover, finding wood supplies, wood exports, net release of carbon from tropical deforestation and tree cover trends. Some relevant discussion on mobilizing for reforestation and the design of a reforestation project. More geared towards commercial demands and social stress on forests.

Rappaport, Bret. 'Weed Laws: A Historical Review and Recommendations'. *Natural Areas Journal*, vol. 12, no. 4, pp. 216–17, 1992. Discussion of weed laws and ordinances that allow the establishment of natural areas on public and private lands while addressing the concerns of neighbours.

Rice, P.F. 'Restoration of a Wildlife Sanctuary in an Urban Setting'. *Journal of Arboriculture*, vol. 17(I), pp. 21–5. Urbana, IL: International Society of Arboriculture, January 1991. Review of the restoration proposals for Cootes Paradise in Hamilton. Briefly discusses implementation techniques and management strategies.

Ruff, Allan R. *Holland and the Ecological Landscapes 1973–1987*. Delft, The Netherlands: Delft University Press, 1987. Since his first book '*Holland and the Ecological Landscape*' that described the successional approach to planting developed by the Dutch, the author has revisited many of the Dutch housing projects he examined in 1979, and presents an analysis of how they have fared over some twenty years.

Schulhof, Richard. 'Public Perceptions of Native Vegetation'. *Restoration and Management Notes*, vol. 7, no. 2, pp. 69–72, Winter 1989. Reviews factors leading to positive public perception based on research carried out at the North Carolina Botanical Garden, Chapel Hill. Public education appears to be the most important factor leading to the appreciation of natural environments. Floral displays (i.e. goldenrod or other wildflowers), water features, novelty (i.e. unusual plant material) and a mystery component to the landscape (i.e., the promise of a new view or different landscape around the corner) were found to be appealing or intriguing.

Sotir, R.B. 'Introduction to Soil Bioengineering Restoration'. In Berger, J.J. (ed.). *Environmental Restoration: Science and Strategies for Restoring the Earth*. Proceedings of conference on restoration, Berkeley, CA, January 1988. Washington DC: Island Press, 1990. Soil bioengineering is an applied science that uses living plant material as a main structural component. It is useful in controlling problems of land instability, where erosion and sedimentation are occurring.

Tanacredi, J.J. 'Management Strategies for Increasing Habitat and Species Diversity in an Urban National Park'. Berger J.J. (ed.). *Environmental Restoration: Science and Strategies for Restoring the Earth*. Proceedings of the conference on environmental restoration, Berkeley, CA, January 1988. Washington DC: Island Press, 1990. Since 1972, restoration programmes have been implemented in the 10,522 hectare Gateway National Recreation Area lands within and adjacent to New York City. Specific activities include active and passive revegetation of impacted sites, restoration of grassland habitat utilized by regionally rare grassland-dependent bird species, creation of fresh water habitat, and a transplant programme to restore populations of native amphibians and reptiles.

ON WILDLIFE

Baines, Chris. *The Wild Side of Town*. London: Elm Tree Books and BBC Publications, 1986. This is a book about urban wildlife and experiencing wild places, the places that wildlife lives in and how people can begin to value them. The author asks, 'Children choose to play in wild, unofficial landscapes, and can't avoid contact with nature. Why do grown ups dismiss these landscapes simply as "untidy"?'. Beautifully and simply written, there is much to learn from this publication from mapping neighbourhood habitat networks, to making habitats and nesting boxes and getting everyone involved.

Baines, C. and J. Smart. *A Guide to Habitat Creation*. London: The Greater London Council, 1984. This book highlights the creation of grasslands (meadows), woodlands and wetlands as urban environments for 'countryside' plants and mammals. It outlines practical techniques for habitat creation in Britain.

Blokpoel, Hans, and Gerard T. Haynes. 'How the Birds took over the Leslie Street Spit'. *Canadian Geographic Journal*, April/May 1978.

Duffey, Eric. 'Effects of Industrial Air Pollution on Wildlife'. *Biology Conservation* no. 15, 1979. This article deals with the little studied effects of air pollutants on the worldwide decline of vertebrate wildlife, and the changes of distribution of certain wildlife species.

Forest Service USDA and Pinchot Institute of Environmental Forestry Research. *Children, Nature, and the Urban environment'. Symposium Proceedings*. USDA Forest Service, General Technical Report, NE-30, USDA, Washington, DC., 1977. Papers on the natural environment and human development, research on urban children and the natural environment, and the community and institutional response to nature education.

Gee, Markus. 'Wild in the City'. *Toronto Globe and Mail*, 12 June 1993. Three publications/reports on the evolution of the Leslie Street Spit in Toronto, reporting on the progress of its evolution, plants and wildlife from 1977 to 1993.

McClanahan, T. R. and R. W. Wolfe. 'Accelerating Forest Succession in a Fragmented Landscape: The Role of Birds and Perches'. *Conservation Biology*, vol. 7, no. 2, pp. 279–86, June 1993. Reviews the effectiveness of bird perches as a tool to accelerate ecological succession. Study of a mined site in central Florida revealed that seedfall beneath perches had a higher diversity of seed genera and seed numbers (up to 150 times higher). While species composition included both early-successional and late-successional species, the harsh conditions of the site favoured the establishment of the early-successional species.

McKeating, Gerald B. (ed.). *Nature and Urban Man Canadian Nature Federation Conference*. Ottawa: Canadian Nature Federation, 1975. A series of papers that examine wildlife in the city, wildlife places, and planning for nature and urban man.

Metropolitan Toronto and Region Conservation Authority, *Thommy Thompson Park, Master Plan and Environmental Assessment*. Addendum report, December 1992.

Noss, Reed F. 'Wildlife Corridors'. In Daniel S. Smith, Paul Cawood Hellmund (eds), *Ecology of Greenways, Design and Function of Linear Conservation Areas*, Minneapolis: University of Minnesota Press, pp. 43–68, n.d. Discussion of wildlife corridors, their function, predation and design issues, including corridor quality and width. Includes guidelines for wildlife corridor design.

Noyes, John H. and Donald R. Progulske (eds). *Symposium on Wildlife in an Urbanizing Environment*. Co-operative Extension Service, University of Massachusetts, Amhurst, June 1974. A series of symposium papers that include philosophy, public and private roles in urban wildlife management, research studies, diseases and people in urban wildlife.

CULTURAL/BEHAVIOUR ISSUES

Francis, Mark. 'Control as a Dimension of Public-Space Quality'. In I. Altman and E. Zube (eds), *Public Spaces and Places*. New York: Plenum, 1989. This paper examines the form and meaning of public space; how it can support public culture and outdoor life, how public spaces be designed and managed to satisfy human needs and expectations, and enable people to control how they use them. Several aspects of user control are examined that include the evolving nature of American public life, and the nature of control as a psychological construct and participation concept.

Thompson, J. William, 'Reconsidering South Central' and 'San Francisco's Gardens of Diversity'. *Landscape Architecture*, vol. 83, no. 1, pp. 46–9 and 60–2, January 1993. Two worthwhile articles on issues of cultural diversity, the social and environmental problems associated with urban unrest in Los Angeles and San Francisco, and the approaches that are being taken to rebuild neighbourhoods through gardens and greening.

Wekerle, Gerda and Carolyn Whitzman. *Safe Cities: Guidelines for Planning, Design and Management*. New York: Van Nostrand Reinhold, 1994. This book addresses problems of violence and fear of crime in cities: where and why crime occurs; what makes some places unsafe and feared; and what politicians, planners, social agencies and communities can do about making them safer. The book stresses the relationship between the physical environment and social problems, and how solutions must be multi-faceted to be effective. Practical solutions and numerous examples of similar experience in many cities in North America and Europe.

ON URBAN FARMING/RECYCLING

American Community Gardening Association. *National Community Gardening Survey* (preliminary report), 1990. This report contains preliminary results of the first year of a two-year survey by the American Community Gardening Association and its members' activities in cities and counties across America. (Information available from American Community Gardening Association (ACGA), 325 Walnut Street, Philadelphia, PA 19106.)

Chowdhury, Tasneem and Christine Furedy. *Urban Sustainability in the Third World: A Review of the Literature*. Winnipeg, Manitoba: The University of Winnipeg, Institute of Urban Studies, 1994. An examination of the literature on issues of poverty, and the relationship of sustainability to development in the context of third world cities. It includes a conceptual analysis of 'sustainable development' that highlights the differences between developed and developing countries; studies that deal with problems of services such as water, sanitation and health and how their absence impacts on the urban poor; urban resources such as urban agriculture and waste management/recycling that might be effectively tapped; planning and policy that addresses the various mechanisms of urban management and the changes needed in international co-operation and assistance.

Creasy, Rosalind. *The Complete Book of Edible Landscaping*. San Francisco: Sierra Club Books, 1982. A useful book that integrates productive plants with design. It deals with pesticide issues, costs, water and soil conservation, history landscape gardening, planning and design for small gardens. Well illustrated with detailed plant layouts and gardening techniques.

European Federation of City Farms. (Guide to city farms in various countries in Europe.) Published by the NFCF Brussels, Belgium (n.d.) Contains history, current activities and addresses of the city farms in Europe, including those in the UK. For further information contact: National Federation of City Farms, Avon Environmental Centre, Junction Road, Brislington, Bristol BS4 3JP, United Kingdom.

Feder, Natasha. *Population and Global Sustainability*. Toronto: Conservation Council of Ontario, May 1992. A summary report on global sustainability in relation to population. Earth's carrying capacity with respect to soil degradation, food production and distribution, forests, and wastes, are examined in addition to other issues such as acid rain, energy family planning and policy. Useful reference guide to world issues.

Freeman, Don. *City of Farmers*. Montreal: McGill/Queens University Press, 1991. A study of urban farmers and farming in Nairobi, Kenya. It covers the context within which the study was made, the nature, background, distribution and practices of the farmers, and the the role and significance of urban agriculture in Nairobi at the levels of the families, the community and the nation.

Furedy, Christine. 'Wastes and the Urban Environment'. *WEDC 12th Conference: Water and Sanitation at Mid-decade*, Calcutta, 1986. This paper examines resource recovery and recycling in Asian societies. Even though recycling of urban wastes occurs extensively in Asian cities, it argues that a fresh concept of the city itself is required - its essential functioning as a natural system.

Memon, Pya Ali and Diana Lee-Smith. 'Urban Agriculture in Kenya'. *CJAS/RCEA* 27:1 1993. This article is an analysis of the characteristics of urban agriculture in Kenya set within a wider conceptual and socio-economic context. It is based on a recent survey by the Mazingira Institute (see Lee-Smith, D. *et al.* 'Urban Food Production and the Cooking Fuel Situation in Urban Kenya'. National Report: Results of a 1985 National Survey. Nairobi: Mazingira Institute, 1987).

Mollison, Bill and David Holmgren. *Permaculture*. Australia: Taragi Publications, vol. 1, 1978 (second publication 1982); vol. 2, 1979. Two well-known books that describe Permaculture as a philosophy of working with, rather than against, nature; an agricultural system that combines landscape design with perennial plants and animals to make a safe and sustainable resource for town and country.

Poerbo, Hasan. 'Urban Solid Waste Management in Bandung: Towards an Integrated Resource Recovery System'. *Environment and Urbanism: Rethinking Local Government; Views from the Third World*, vol. 3 no. 1, pp. 60–9, April 1991. This paper explores the potential for developing an integrated resources and recovery system in association with scavengers in Bandung, Indonesia. The goal was to make the scavengers the centre of a decentralized and technology-independent recycling and waste management system based on modules managed by different groups throughout the city. The paper describes the development of linkages and relationships between scavengers and researchers. This work has been conducted at a small scale. An incremental approach is proposed that would lead to the creation of a recovery system for a large urban centre.

Sachs, Ignacy and Dana Silk. *Food and Energy: Strategies for Sustainable Development*. Paris: United Nations University, 1989. Publication based on the activities of the Food–Energy Nexus Program (FEN) of the United Nations which took place between 1983 and 1988. Discusses the contribution urban agriculture can make using the survival skills of rural refugees to grow food, its contribution to the 'hungry season' between major harvests, types of crops and obstacles to growing food gardens.

Smit, Jac and Joe Nasr. 'Urban Agriculture for Sustainable Cities: Using Wastes and Idle Land and Water Bodies as Resources'. *Environment and Urbanization*, vol. 4, no. 2, October 1992. The paper describes how cities can be transformed from being only consumers of food and agricultural products into important resource-conserving, health-improving, sustainable generators of these products.

Tinker, Irene (ed.). 'Urban Food Production – Neglected Resource for Food and Jobs'. *Hunger Notes*, vol. 18 no: 2, Autumn 1992. A series of articles on food production in cities including Africa (Kampala) and the United States (San Francisco).

ON CLIMATE/ENERGY

Back, Boudewijn and Norman Pressman. *Climate Sensitive Urban Space: Concepts and Tools for Humanizing Cities*. Delft: Colophon, 1992. A practical book on environmentally sensitive approaches to the design of urban spaces with climate in mind. Includes the social significance of public spaces, mobility and liveability, historic precedents, traffic calming, improvements for pedestrians and cyclists and recommendations for wind and solar design in hot and cold climates.

McPherson, E. Gregory. 'Economic Modeling for Large-scale Tree Plantings'. *Proceedings of the ACEEE 1990 Summer Study on Energy Efficiency in Buildings*, vol. 4, American Council for an Energy-Efficient Economy, Washington DC. Large-scale urban tree planting is advocated to conserve energy and improve environmental quality, yet little data exist to evaluate its economic and ecologic implications. This paper describes an economic-ecologic model applied to the Trees for Tucson/Global reLeaf reforestation programme.

McPherson, E. Gregory, James Simpson and Margaret Livingston. 'Effects of Three Landscape Treatments on Residential Energy and Water use in Tucson, Arizona'. *Energy and Buildings*, no. 13: 127–38, Elsevier Sequoia, 1989. This paper reports on research on the role of vegetation in reducing the cooling loads of buildings in hot arid climates. Three similar model buildings were surrounded by different types of landscapes: turf, rock mulch with shrub foundation planting and rock mulch with no planting. Irrigation water use and electricity required to power air conditioners and interior lights were measured for two week long periods, and energy use measured. Preliminary findings suggest that localized effects of vegetation on building microclimate may be more significant than boundary layer effects in hot arid regions.

McPherson, E. Gregory, David J. Nowak, and Rowan A. Rowntree. *Chicago's Evolving Urban Forest Ecosystem: Results of the Chicago Urban Forest Climate Project*. USDA, Forest Service, Northeastern Forest Experiment Station, General Technical Report NE-169, 1993. This highly informative publication includes the role of vegetation in the city, urban forest structure, the influence on wind and air temperatures, air pollution and atmospheric carbon reduction and energy saving potential of trees.

Manty, Jorma and Norman Pressman (eds). *Cities Designed for Winter*. Helsinki: Building Books Ltd, 1988. Paints a broad picture of politics, planning, and design issues for northern cities. The book highlights the essential concerns of winter cities and documents useful experience by international experts. Case studies from China, Finland, Norway, Canada and other countries with similar climates. An important document for local authorities, urban analysts, designers and town planners, and concerned citizens.

Oke, T.R. 'The Significance of the Atmosphere in Planning Human Settlements'. In Wiken, E.B. and G. Ironside (eds), *Ecological (Bio-physical) Land Classification in Urban Areas*. Ecological Land Classification Series No. 3. Environment Canada, Ottawa, 1977. This paper outlines the importance of incorporating atmospheric considerations in the design of settlements. It stresses the interaction between the atmosphere and the built environment. The regional climate imposes constraints upon the nature of the settlement, and the nature of the settlement evokes a change in the local climate. Some of the fundamental aspects of these interactions are examined and how they can be incorporated into the design of urban areas.

Paehlke, Robert. *The Environmental Effects of Urban Intensification*. Report prepared for Municipal Planning Policy Branch, Ministry of Municipal Affairs, Government of Ontario, Canada, March 1991. This report deals with urban sprawl as a key environmental concern in Ontario that has resulted in loss of agricultural land, damage to habitat and biological diversity. The dominant role of the automobile in urbanization is associated with high energy uses, and air quality problems. The report examines the positive as well as potential negative impacts of intensification in ameliorating urban problems. Very useful information on energy use, air quality and environmental issues in relation to populations and urban form. Extensive annotated bibliography.

Philleo, Jerryne. *The Davis Energy Book*. City of Davis, March 1981. This publication describes the development of an energy conservation strategy for the City of Davis, California that included public involvement, climate research, energy conservation building code and city conservation plans, transportation, trees and related energy policy issues. There have subsequently been numerous changes and improvements to these policies.

INDEX

•

Bold page numbers indicate illustrations and figures; *italics* indicate tables.